Rainbows, Snowflakes, and Quarks

1 2 3 4 5 6 7 8 9   FGR FGR   8 7 6 5 4

ISBN 0-07-067545-7

LIBRARY OF CONGRESS CATALOGING IN PUBLICATION DATA

Von Baeyer, Hans C.
Rainbows, snowflakes, and quarks.
Bibliography: p.
1. Physics—Popular works.   I. Title.
QC24.5.V67   1984   530   84–939
ISBN 0–07–067545–7

Book design by Sharkey Design

Illustrations by Laura Hartman

Rainbows, Snowflakes, and Quarks

Physics and the World Around Us

Hans C. von Baeyer

McGraw-Hill Book Company

New York   St. Louis   San Francisco   Hamburg   Mexico   Toronto

For Melissa, Christopher, and Barbara

Contents

Rainbows, Snowflakes, and Quarks

Introduction:
The Nature
of Physics

The word *physics*, translated literally from the Greek, means "natural things." It refers to rainbows and snowflakes, clouds and lightning, waterfalls and whirlpools, rosy dawns and brilliant sunsets, ocean surfs and ripples on puddles, to the multitude of splendid forms and ceaseless transformations that we experience as the material world.

It is unfortunate that "physics" in English sounds so clinical. The harsh sibilance of the word suggests a cerebral enterprise of cold facts, incomprehensible calculations, and peculiar devices remote from the daily affairs of ordinary people, connotations that physics shares with all the sciences. The humanities, by contrast, imply by their very name an involvement with human concerns and hence an easy accessibility. Music, art, and poetry are appreciated as food for the soul, but physics often faces aversion mingled with fear of the unknown. A return to its roots, a fresh look at its real meaning, removes some of the estrangement and brings physics closer to home.

Physics began in ancient Greece with a study of natural phenomena. Since then the world has changed, but the phenomena endure. The dew sparkles now just as it did then, its flashing colors as delightful to watch and its formation as fascinating to explore today as three thousand years ago. The phenomena have not only retained their appearances, but they also remain at the center of physics. A complete explanation of the formation of dew is every bit as sophisticated as the laws of nature that govern lasers and black holes, but dew is more accessible even in the space age. The essence of physics can be found nearby, even though it is often buried beneath the surface of things.

The aim of this book is to bring physics closer to the humanities by examining some familiar experiences of nature. Breathless expositions of the marvels of high technology and the mysteries of theoretical physics yield the stage to homely phenomena, made homelier by close examination. Only the last chapter touches briefly on one of the more abstract ideas of modern physics. Continuity with the past is valued more than novelty.

Theoretical physics has become sophisticated, but the ways of thinking and the underlying assumptions have not. A handful of central ideas have remained remarkably unchanged over the centuries. These essential ideas have been called *themata* (plural of *thema* or "theme") by Gerald Holton, who has traced their influence on the work of scientists throughout history. Themata are unspoken premises, prejudices, assumptions based on intuition, they are persistent motifs or subconscious biases that guide the thinking of even the most objective scientists. If the experimental basis and the theoretical analysis are stripped from physics, a residue of unfounded principles remains; these are themata.

Among the persistent themata are the hypothesis that matter consists of discrete atoms, and the contrasting view of the world as a continuum. These two principles strive for

Introduction: The Nature of Physics

ascendancy even today and introduce an essential tension to modern physics. Other themata are the continual search for symmetry since Plato, the preoccupation with integers inspired by Pythagoras, and the insistence on parsimony in enumerating axioms that is associated with the name of Euclid. There are, furthermore, the realization that quantification must precede analysis, the ancient idea that some physical quantities are conserved, the use of statistical and probabilistic approaches to the understanding of many phenomena, the focus on the role of the observer's point of view that culminates in relativity, the value placed on generality and on simplicity in explanation, and the concept of causality. Perhaps the most fundamental thema is the belief in the possibility of a single consistent mathematical description of nature.

Themata clearly do not grow out of experimental observation of phenomena, nor are they explicit theoretical assumptions. Yet they shape not only the answers, but the very questions formulated by physicists. They are not part of the usual canon of scientific jargon, and perhaps because of that they are as accessible to laymen as the natural phenomena themselves. Like the phenomena, they link the thoughts of scientists through the ages. James Clerk Maxwell, who a hundred years ago devised a mathematical analysis of electricity and magnetism in a language that is as foreign as Greek to most people, recognized his connection with previous generations through common themata and a universal sense of wonder:

Is space infinite, and in what sense? Is the material world infinite in extent, and are all places within that extent equally full of matter? Do atoms exist, or is matter infinitely divisible? The discussion of questions of this kind has been going on ever since men began to reason, and to each of us, as soon as we obtain the use of our faculties, the same old questions arise as fresh as ever. They form as essential a part of science of the nineteenth century of our era, as that of the fifth century before it.

Of the three elements of physics—phenomena, themata, and analysis—the first two are both ancient and simple. Only the third is technical, difficult, and very abstract. It is unfortunate and ironic that much popularization of science focuses on theoretical analysis, attempting to make it palatable by simplification and circumlocution. This approach misses the comforting fact that the motivations and tacit assumptions—the themata—of the modern physicist closely resemble those of the Greek philosopher, the medieval scholar, and the Renaissance thinker. Maxwell's questions, for example, are of real and immediate concern to the twentieth-century cosmologist and high-energy theorist. Understanding the historical unity and continuity of physics does much to humanize it.

In the end, however, analysis must be faced because it is as essential to physics as are phenomena and themata. Here a formidable barrier to popular understanding arises: the mathematical language in which physics is written. Translation of equations into words is awkward, but no more so than translation of poetry into other languages. The real problem is the extreme economy of mathematical expression, which crystallizes complex ideas into almost irreducibly meager sets of symbols. Newton writes $F = ma$ and thereby captures a whole universe of mechanical interactions. This way of expressing ideas contrasts sharply with the method of the humanities. Shakespeare cannot condense the totality of his thoughts and feelings about Lear into a sentence like "Lear is mad." Instead he must write the play in all its redundancy, ambiguity, vagueness, verbosity, and obscurity. The ultimate message of *Lear* may be brief, but it cannot be reduced to a few words. The messy wildness of Shakespeare's play comes closer than Newton's clarity to the way people think and feel. Mathematics is just too surgically precise.

But the comparison is misleading. Hidden behind the equation $F = ma$ are the definitions of the symbols, the philosophical problems of their interpretation, the historical antecedents of the theory and its complicated, imprecise and ambiguous experimental tests, its realms of applicability and limitations,

Introduction: The Nature of Physics

its practical consequences, its equivalent formulations—in short, the meaning. The physicist who appreciates the beautiful conciseness of the equation is aware of the vast messy world of physics in the background without which the four little symbols would be senseless. He is cheating when he pretends that F = ma tells the whole story, just as the scholar is cheating when he substitutes a synopsis of *Lear* for the play.

Mathematics is a compact language that means a lot to those who are able to interpret and use the symbols, but nothing to those who are not. For this reason popularizers of physics must endeavor to supply the meaning that is not apparent in the mathematical descriptions. Introductory textbooks try different ways of fleshing out the bare mathematical laws of physics. Some emphasize the history of science, some stress useful applications in everyday life, some rely on anecdotes about people and events, some try to stimulate readers to make their own discoveries. Each of these attempts is valid, but a mixture is more effective because physics involves all of them, and because different students respond to different approaches. Similarly, *Lear* is not exhausted by a study of its organization, by historical references, by psychological analysis, or by a glossary of unfamiliar words. An effective editorial apparatus includes the most significant portions of all four. Ultimately, however, the literary editor has an easier task than the scientific popularizer because *Lear*, even without footnotes, speaks to readers in a way that F = ma does not.

The most powerful device for the translation of physics from mathematical language into words is analogy. In some general sense all learning is analogical because it couches unfamiliar concepts in familiar images and terms. In this case, analogy is especially suitable because it is one of the favorite tools among professional scientists. For example, by virtue of the unexpected analogy to the solar system, Niels Bohr was able to understand the hydrogen atom. A few years later, the similarity between the vibrations of a violin string and the oscillations of electrons in the hydrogen atom enabled Erwin Schrödinger to refine and extend Bohr's theory. Since advances in theoretical physics often contain analogies, the physicist is familiar with the technique and adept at using it as a bridge for the uninitiated to his arcane science.

Closely related to analogy in the vocabulary of physics is the word model. The mechanical models of past generations have given way to mathematical models, but the idea is the same. A scientific model is an artificial construct that serves as analog for a natural phenomenon. It is simpler, better understood and, like the scale model of a ship, more easily manipulated than the original. It can be adjusted and refined until its properties mirror the empirical observations with sufficient accuracy to give confidence in its predictive power.

Models played a particularly important role in physics during the last part of the nineteenth century when the universe was thought of as a vast, exquisitely tuned clockwork. Lord Kelvin, one of the greatest physicists of the time, maintained that he could not understand any phenomenon until he had imagined a mechanical model of it. The compulsion to build mechanical models reached its height around the turn of the century when Joseph Larmor proposed a representation of the vacuum as an immense aggregate of tiny interconnected gyroscopes. They were necessary, Larmor thought, to mimic the observed properties of the vacuum that seemed paradoxical at the time. Einstein boldly swept away the creaking machinery in 1905 when he declared the vacuum to be empty, but since then theoreticians have been busy filling it up again with more ethereal but no less bizarre constructions.

For the popularizer of science, the models in the attic of physics are wonderfully handy. Stripped of all claims to reality, they now serve as analogies to make modern theories more picturesque and comprehensible. Electricity, for example, was once thought of as a fluid. This model has long since been discarded, but the analogy with water currents, which gives rise to the application of such words as *current, flow,* and *capacity* to electricity, is still powerfully suggestive. In astronomy, mechanical models of the solar system no longer represent the forces that keep the planets in orbit, but they are valuable teaching devices. Even Bohr's model of the hydrogen atom, which seemed so real in 1913, is today regarded as a mere historical curiosity, but as an analogy it continues to introduce younger generations to quantum mechanics. That a stylized picture of Bohr's planetary atom adorns the letterheads of countless small hi-tech companies attests to its undiminished appeal. The physicist's model of last year has today become a universal pictograph.

In the humanities, analogies and models are called myths; they serve, as they do in science, to make the incomprehensible comprehensible. In this sense, Genesis is an analogy for the creation of the world. As long as it is understood to represent a reality that is very different and much more complicated, it retains its affective validity. Only when it is mistaken for the actual story does it become misleading and harmful.

More generally, religion is the greatest analogy of all. The scientist, aware of the unfathomable complexity of nature and accustomed to both the explanatory power of analogy and its limitations, turns to religion with less hesitation than popular opinion, unintentionally or intentionally misguided, would suppose.

For Sir Isaac Newton, God was not a myth but a real and necessary component of the universe. Beginning with the declaration "This most beautiful system of the sun, planets, and comets could only proceed from the counsel and dominion of an intelligent and powerful Being," Newton discussed at length what can, and what cannot, be deduced about God from an examination of natural phenomena. He firmly asserted the legitimacy of this inquiry, which forms part of his *Mathematical Principles of Natural Philosophy:* "And this much concerning God; to discourse of whom from the appearance of things, does certainly belong to natural philosophy." A quarter of a millennium later, Einstein wrote in his intellectual autobiography: "[The scientist's] religious feeling takes the form of a rapturous amazement at the harmony of natural law, which reveals an intelligence of such superiority that, compared to it, all the systematic thinking and acting of human beings is an utterly insignificant reflection." God has become the human representation, the analog, of the ineffable intelligence that permeates nature. To the extent that the scientist senses the beauty and order of the universe, he is religious. Halfway in time between Newton and Einstein, Goethe summed up the analogical relation between science and religion in a typically ambiguous maxim: "He who possesses science and art also has religion; but he who possesses neither of those two, let him have religion!"

This book is written for people who have art, and would add a little science. It dwells on phenomena and on some of the people who have tried to understand them. It picks out, in the rich tapestry of the history of science, traces of those golden strands, the themata, that hold the whole fabric together. It makes reference every now and then, playfully and gingerly, to mathematical analysis translated into words by means of analogies and models. It endeavors above all to communicate a little of the sense of wonder bordering on reverence that nature inspires in scientists.

Motion

Children romping at the beach, tossing a ball into the summer sky; a home run over the fence on a golden prairie evening, the sun's last rays stretching tall shadows across the diamond; a long forward pass through the brittle autumn air to the twenty-yard line; a high lob over the net to the tennis player racing to the back of the court; a free throw in the hushed expectancy of the high-school gymnasium—from this profusion of images a common element captures the interest of the physicist: the flight of the ball. Fast enough to require close attention but slow enough to stop the heart, inevitable in its course but unpredictable in its effect, smooth but not uniform, the same motion was followed by rocks hurled out of prehistoric volcanoes and will be traced by projectiles when all the stadiums have crumbled.

Physicists seek out those features of inanimate nature that are simple and universal. This is the secret of the prodi-

gious success of physical science. It is not so much a matter
of finding answers as it is of asking the right questions. The
trick is keeping the questions simple and sticking to universal
phenomena. Physicists don't ask BIG questions: What is life?
How do you make love stay? How do you cure cancer? What
stops inflation? Even within the discipline itself, such complicated or unique phenomena as whirlpools and the birth of
the earth lead to uncertain explanations. But simple questions—How fast does a stone fall? How do billiard balls col-
lide?—are answered unambiguously by careful observation of
repeated experiments and by logical thinking. One such question of universal appeal concerns the path of projectiles, the
flight of balls and bullets. Of all the questions that might occur
to the watcher of a ball game, it is not speculation about the
role of play in society, nor the physiology of throwing, nor

the human need to succeed, nor the choreography of the limbs of the players, but the description of the ball's path that engages the physicist. Perhaps it is the easiest problem to solve. Certainly it concerns, as Galileo Galilei maintained, one of the most basic effects in science.

Like rocketry and atomic energy, the modern science of projectile motion had its beginning in military application. Its inventor was Niccolò Tartaglia, an eminent mathematician born in 1500, a hundred years before Galileo made mechanics into a systematic science. Tartaglia owes his name, which means "stutter," to the stroke of a sword that split his chin at an early age and caused a permanent speech impediment. He became interested in the flight of projectiles when a soldier asked him for the angle of elevation at which a cannon would achieve its greatest range. Tartaglia's correct theoretical answer of 45° surprised the experts; they thought it should be smaller. A test, appropriately enlivened by bets on the side, confirmed the mathematical prediction and prompted Tartaglia to delve further into the subject. By 1532 his notes had grown into a treatise, but he refrained from publication. The reason for his diffidence is highly creditable: He felt that it would be immoral to use science to help Christians slaughter Christians more efficiently. His decision was a rare example of what some people feel is responsible behavior for scientists. Then, as now, publication was the means of achieving fame and fortune, so Tartaglia's renunciation of publication represented a costly sacrifice. Alas, his scruples were short-lived. In 1537 Venice had reason to fear invasion by the infidel Turks and in the interest of internal security Tartaglia published his book, the first scientific text on ballistics. Then, as now, war against another religion seemed more justified than war against one's brothers. Or perhaps it was a question of self-defense.

Because cannonballs, and even beach balls, travel so fast, it is difficult for the unaided eye to trace their paths. In fact, the trajectory is an imaginary line devoid of tangible markers.

It is not obvious how a theoretical trajectory, calculated and drawn on paper, can be compared with experiment. Over the centuries a variety of methods for performing the test have been invented. In Florence, tucked away behind the Uffizi and ignored by an ocean of tourists surging around the immortal works of art, a little museum called Museo Nazionale di Storia della Scienze stands as a monument to rational culture. In the back room, beyond Galileo's finger displayed in a silver monstrance, a collection of mechanical teaching devices includes a wooden catapult followed by a downward sloping sequence of adjustable hoops. When all the hoops are perfectly centered on the correct trajectory, a ball, after being launched, will follow its path unimpeded to the floor. However, if one of the hoops is out of place, the ball will hit it. Changes in the angle and speed of launch must be accompanied by adjustments in the positions of the hoops. In this way the trajectory is mapped out by the centers of the hoops and can be studied, drawn, and measured at leisure. Today, mechanical devices are replaced by stroboscopic lights or slow-motion film that freeze the action.

But there is a simpler way to see the whole path at once. A stream of water discharged into the air and continuing unhindered defines a trajectory. Each drop can be thought of as an individual projectile, its path made visible by all the others that precede and follow along the same route. A garden hose provides the most versatile demonstration, a drinking fountain the most ubiquitous. In a fountain, the path of projectiles is made visible not by stopping the motion, but rather by repeating it ceaselessly. As each drop moves from a spot, its place is taken by the next one. The problem of defining the shape of the trajectory of a cannonball is replaced by the problem of defining the shape of the arch formed by the stream of water from a fountain.

Galileo was the first to describe the exact geometry of that shape and to derive it from the fundamental principles of mechanics. He was well aware of the significance of the discovery. In his last and supreme work, *Dialogues Concerning Two New Sciences,* published when he was seventy-four, he noted that since ancient times "it has been observed that missiles or projectiles trace out a line somehow curved," but he proudly announced that he was the first to demonstrate just what that curve is. Nevertheless, he realized that the importance of his contribution lay not so much in the detail as in the method, by which, he hoped, "there will be opened a gateway and a road to a large and excellent science into which minds more piercing than mine shall penetrate to recesses still deeper."

He was right. His derivation of the ballistic trajectory opened the road that led from Newton, who was born in the year that Galileo died, to Einstein in this century. It is the link between mathematics and nature, exemplified in the lowly drinking fountain, that characterizes modern science.

The most obvious guess about that shape, and most natural to the Greek philosophers who considered straight lines and circles to be divine and perfect, is that the trajectories of balls, bullets, and water drops are parts of circles. A complication immediately arises in the case of an object that is launched almost straight up. The trajectory seems to consist of three distinct parts: an initial straight leg, a curved portion that might be a semicircle, and a third straight section symmetric to the first. Galileo, after years of thinking and innumerable false starts, found one simple mathematical formula that describes the trajectory in all cases. The formula corresponds to a geometrical shape with a familiar name but an unfamiliar definition: the parabola.

The parabola itself, as a mathematical concept, is much older than Galileo. Along with the hyperbola and the ellipse, it was introduced in the book *Conics* by Apollonius of Perga in the third century B.C. Imagine a cone like a dunce cap. If a sharp knife slices straight down through the side of the cap, cutting off a piece, the edge of the cap along the cut is a hyperbola. If, instead, the cut is slanted parallel to the side of the cap, a parabola is carved out. If the cut is almost but not quite horizontal, an ellipse results. The least interesting cut is exactly horizontal, yielding a circular edge.

The names *hyperbola, parabola,* and *ellipse* are derived from words meaning "to exceed," "to equal," and "to fall short of." They refer to the slope of the cut that exceeds, equals, or falls short of the slope of the cone. The same roots lead to hyperbole, parable, and ellipsis, referring to descriptions that exceed the original, equal it, or fall short by omission. Thus the jargon of English grammar echoes faintly with the names of the crown jewels of Greek geometry—hyperbola, parabola, ellipse.

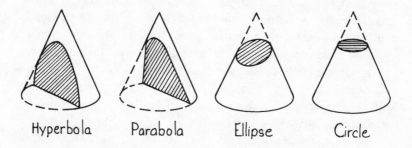

Hyperbola   Parabola   Ellipse   Circle

Apollonius worked out the mathematical formulas for describing conic sections. The simplest, in modern terms, is that of the parabola. If a parabola is drawn with its opening downward and its vertex at the origin of a coordinate system, then for any point on the parabola the vertical distance from the origin is proportional to the square of the horizontal distance. This means that if a point sliding down along the parabola moves two units to the right, it simultaneously moves four units straight down.

To demonstrate that the path of a projectile is a parabola, Galileo needed three ideas, which he worked out with the help of logic and experiment in a lifelong struggle against Aristotelian authority. They are the law of free fall, the law of inertia, and the principle of relativity. These rules of mechanics are exemplified in such trifles of nature as the shape of the jet of a drinking fountain.

Galileo's law of free fall is the rule that an object falling from rest covers distances in equal intervals of time that bear the proportions $1:3:5:7:9$. . . . If, for example, a stone dropped into a well passes one rung of a ladder during the first heartbeat, it will pass three rungs during the next, five after that, then seven, and so on. How Galileo arrived at this elegant and suggestive result is not clear, but we know that he corroborated it experimentally. Perhaps he noticed a remarkable quality displayed only by the sequence of odd numbers. Suppose that in the example of the well the basic interval of time is chosen to be two heartbeats instead of one. Then successive distances fallen are $1 + 3 = 4$ rungs, $5 + 7 = 12$ rungs, $9 + 11 = 20$ rungs, and so forth. The distances 4, 12, 20 . . ., bear the ratios $1:3:5$ . . . to each other, as is required by Galileo's law of free fall. Other sequences, such as $1:2:3:4$ . . . and $2:4:6:8$ . . ., do not have this nice property that leaves the ratios of successive distances the same, independent of the length of the time interval.

From the law of free fall, the total distance of an object from its starting point can be derived. After one heartbeat that distance is 1 rung. After two heartbeats it is $1 + 3 = 4$ rungs. After three it is $1 + 3 + 5 = 9$ rungs, and so forth. The resulting sequence of numbers is, *mirabile dictu,* the list of squares of integers. To summarize: *During* successive intervals of time, the distances fallen are to each other as $1:3:5:7$. . . . *After* successive intervals of time, the total distances from the starting point are as $1:4:9:16$. . . . The followers of Pythagoras, who believed that the secrets of nature are encoded in numerological relationships, would have been delighted by these rules.

In order to verify them experimentally, which is difficult for objects falling vertically, Galileo used an ingenious device to slow gravity. Instead of dropping marbles, he let them roll

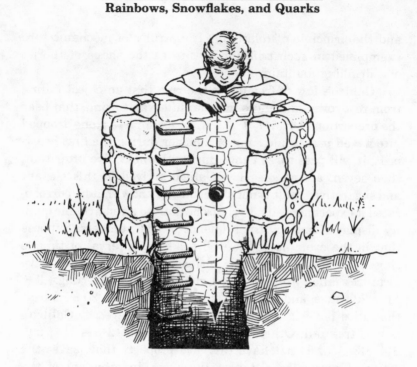

on long slanted inclines, stretching the distances and increasing the time intervals for greater ease of measurement.

Galileo's second tool was the law of inertia. This idea is so simple to state that an intellectual effort is required to appreciate the struggle of its development through the centuries. Even Galileo didn't see the law in its full stark simplicity. Conforming to the observation that ontogeny often recapitulates phylogeny, modern freshmen also find it a struggle. The law of inertia is this: In the absence of external forces, a moving object will continue to move forever. A hammer thrown from a capsule in outer space will never stop. A hockey puck on ideal frictionless ice will skitter off to infinity without slowing down. A perfect car on a perfectly flat and level road with no air resistance will roll on indefinitely.

Neither the Greeks nor the scholastic philosophers discovered the law of inertia. Intuition, informed by everyday observation, seems to contradict it. In the absence of external forces, moving objects stop: A cart moves while it is pushed, but when the motive power is removed, it quickly comes to a halt. In order to formulate the law of inertia, friction and air resistance must be imagined away, but the degree of abstraction needed to ignore them is surprisingly high.

Galileo achieved the requisite detachment by means of a thought experiment. He argued as follows: On a smooth rising slope, a rolling marble will lose speed. On a smooth declining slope, a rolling marble will gain speed. What will happen on a smooth level plane? The moving marble will neither slow down nor accelerate, and must therefore retain its original speed. That is the law of inertia. It has become famous as Newton's first law of motion, even though it should bear Galileo's name.

The third element of Galileo's derivation of the parabolic trajectory is the principle of relativity. It was not called by that name until much later, but his statement and use of the idea are almost unchanged to this day. Galileo realized that projectile motion is complicated by the fact that a cannonball moves forward and downward at the same time, so the question arises: How do these two motions affect each other? The answer is as simple as it is surprising: Not at all. The time it takes an object to fall ten feet, and the time it takes a cannonball to fly two miles while dropping ten feet, are precisely the same. Although intuition recoils at the claim that a bullet shot horizontally and the shell simultaneously ejected from the gun will hit the ground together, observation bears it out. The horizontal motion is simply superimposed on the vertical motion without changing it.

To prove this proposition, Galileo invites us to consider a ball dropped by a sailor perched on the mast of a ship. If the ship is in the harbor, the ball will hit the deck after a certain amount of time. If the ship is moving forward with uniform speed, Galileo claims, the ball will hit the deck in

the same spot after the same interval of time, even though ship, mast, sailor and ball have been moving forward all along.

An observer on the moving ship will see the ball fall in a straight line. An observer on the shore, however, will see the ball follow a path that is curved forward in the direction of the ship's motion. The description of the trajectory is *relative* to the situation of the observer.

The most eloquent illustration of the idea that forward motion does not affect downward motion occurs on the second day of the *Dialogue Concerning the Two Chief World Systems,* Galileo's defense of Copernicus. Here, in order to escape the troublesome effect of air resistance on cannonballs, Galileo takes us below.

Shut yourself up with some friend in the main cabin below decks on some large ship, and have with you there some flies, butterflies, and other small flying animals. Have a large bowl of water with some fish in it; hang up a bottle that empties drop by drop into a wide vessel beneath it. With the ship standing still, observe carefully how the little animals fly with equal speed to all sides of the cabin. The fish swim indifferently in all directions; the drops fall into the vessel beneath; and, in throwing something to your friend, you need throw it no more strongly in one direction than another, the distances being equal; jumping with your feet together, you pass equal spaces in every direction. When you have observed all these things carefully (though there is no doubt that when the ship is standing still everything must happen in this way), have the ship proceed with any speed you like, so long as the motion is uniform and not fluctuating this way and that. You will discover not the least change in all the effects named, nor could you tell from any of them whether the ship was moving or standing still. In jumping, you will pass on the floor the same spaces as before, nor will you make larger jumps toward the stern than toward the prow even though the ship is moving quite rapidly, despite the fact that during the time that you are in the air the floor under you will be going in a direction opposite to your jump. In throwing something to your companion, you will need no more force to get it to him whether he is in the direction of the bow or the stern, with yourself situated opposite. The droplets will fall as before into the vessel beneath without dropping toward the stern, although while the drops are in the air the ship runs many spans.

How powerfully this story convinces us, compared to the dull words of modern writers! Consider this passage from a lecture by Henri Poincaré entitled *The Principles of Mathematical Physics,* delivered in St. Louis in 1904:

The principle of relativity [is that] according to which the laws of physical phenomena should be the same, whether for an observer fixed, or for an observer carried along in a uniform movement of translation; so that we have not and could not have any means of discerning whether or not we are carried along in such a motion.

The content of both statements is precisely the same. The most famous use of the principle of relativity came the next year, when Albert Einstein repeated it once more and built upon it a whole new world picture known as the "Theory of Relativity."

The pieces are now in place for the analysis of a trajectory. The simplest case is that of a cannonball shot horizontally from atop a high cliff. Except for air resistance, which shall be ignored here, the forward motion of the projectile is unimpeded. According to the law of inertia, the ball will continue to move forward without slowing down or speeding up. It covers a horizontal distance proportional to the number of time intervals. Simultaneously and independently, according to the principle of relativity, the ball is falling. The law of free fall decrees that the vertical distance from the top of the cliff is proportional to the *square* of the number of time intervals. It follows that the vertical distance is proportional to the square of the horizontal distance. This relationship corresponds exactly to Apollonius' formula for a parabola. Hence the trajectory is a parabola.

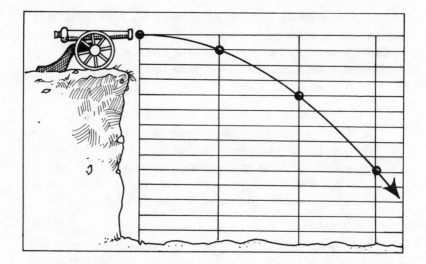

Motion

The problem of motion, which had defied analysis since Parmenides' efforts in 500 B.C., has finally yielded to geometry. Almost two millennia of intellectual struggle culminate in the remark that the stream from a drinking fountain forms a parabola.

The parabola was not the only conic section that was suddenly found to be manifested in nature. At almost the same time as Galileo's studies, Johannes Kepler discovered that the planets move not in circular orbits but along elliptical paths. The Greek preoccupation with perfect circles had so completely dominated astronomy that any other pattern had been simply inconceivable. Even the radical innovator Copernicus, who dared to assume that the earth moves, had constructed his celestial model with circles inscribed on crystal spheres. Kepler himself had begun his investigation of planetary motion with nested circles, and even Galileo could not bring himself to abandon circles in astronomy. Spheres and circles are perfectly symmetric, in the sense that they look the same from all directions. In comparison the ellipses and parabolas that took their place, while symmetric in some directions, are ugly lopsided constructions. With their introduction it appeared that divine perfection had receded into the realm of philosophy and theology, while physics began to descend to worldly imperfection and impermanence.

But the concept of symmetry is not only ancient and powerful, it is also persistent. Whenever it disappears from view it reappears at a deeper level of analysis. In the case of mechanics, spherical symmetry was restored soon after Galileo's death. Isaac Newton's law of universal gravitation and his law of motion combine into a little equation containing a half-dozen symbols. To the mathematician this equation displays perfect and obvious spherical symmetry. Newton derived from it the prediction that any object moving under the influence of gravity must follow an ellipse, a hyperbola, a parabola, or a circle. Thus he brought firmly into the realm of physics the conics that Galileo and Kepler had so reluctantly borrowed from geometry.

How is it possible that Newton's symmetrical equation can produce ellipses with a lower degree of symmetry? When the Greeks focused on circular celestial motion, they did not look deeply enough into nature. The underlying principle is not the symmetry of the sphere, but that of the cone. The cone itself is symmetric, but when cut it yields a variety of conic sections whose lack of complete symmetry is an accidental consequence of the slope of the knife. Similarly, a symmetrical equation can have unsymmetrical solutions depending on the initial conditions. A physical example of this mathematical statement is provided by an artificial satellite. Launched near the earth it can, depending on the speed and direction of launch, follow a subsequent path that is any one of the four conic sections, even though the earth is round and gravity is distributed symmetrically about it. Modern physicists are just as intrigued and delighted by the subtle symmetry of their equations as the Greek philosophers were by the more obvious symmetry of the circular planetary orbits they assumed.

Symmetry is restored to the path of a ball through the summer air, but a mystery remains. Why are the curves that sprang from the mind of Apollonius of Perga reproduced by the trajectories of bullets and planets? What is it that allows a congruence between the pure invention of a mathematician and the course of a natural phenomenon? Why does mathematics describe the real world? It is easy to imagine a universe in which this is not true. If, for example, God chose to manifest His will directly, instead of through the mediation of natural laws, then physics would be a different science. Apples would sometimes fall fast, sometimes slowly, sometimes not at all, according to the whim of the Almighty. The day's length would vary according to His mood. Light would bend to His will. Mathematical laws would be useless because they would not incorporate the heavenly desire. But in fact mathematics works. The discoveries that an ellipse describes the orbit of Mars and a parabola a fountain are, *a priori,* surprises. There must be something hidden here, some powerful unknown connection between mathematics and nature. What is it?

Part of the answer may be that to some extent we force our ideas on nature. A trajectory, after all, is not perfectly parabolic, nor a planetary orbit perfectly elliptical. We choose to ignore the imperfections and concentrate on finding in nature what our minds have invented. If computers had been developed before geometry and calculus, the descriptions of trajectories and orbits would consist of long lists of numbers instead of conic sections, more accurate but less surprising. Still, the basic question would remain. Instead of wondering about the connection between nature and geometry, we would be led to ask: What is the connection between nature and arithmetic?

Another part of the answer may be that without realizing it, Apollonius, Galileo, and Kepler were exploring the same problem—the structure of space and time. If gravity, like motion, is a property of space and time, then the same patterns should reveal themselves in both mathematical and physical inquiries. Conic sections should crop up in geometry and in trajectories of objects in the grip of gravity. But in that case, why don't *all* curves invented by mathematicians have counterparts in nature?

For followers of Plato like Johannes Kepler, the answer is that nature makes use of a certain small number of forms or patterns in all its designs. In this view, parabolas and ellipses pre-existed in the mind of Apollonius and he had only to uncover them, in the same way that Galileo uncovered the parabola in the drinking fountain and Kepler the ellipses in the solar system.

But these are speculations, and of no great concern to physicists. That mathematics is the language of the book of nature is one of the fundamental unquestioned assumptions, one of the themata, of physics. Einstein neatly dismissed the whole problem when he said: "The most incomprehensible thing about the world is that it is comprehensible."

Gravity

On the fourteenth of March, 1979, the centenary of Albert Einstein's birth was observed throughout the world. It was a jubilee of pure reason and a memorial to human kindness, humility, and decency. For over half a century Einstein's name has been a household word far beyond the boundaries of the community of physicists. He is a folk hero, a somewhat mysterious genius whose work is incomprehensible to the uninitiated but whose pronouncements on every subject command respectful attention. His birthday refreshed and enhanced this almost mythical reputation. His familiar gentle face topped by a wild shock of white hair sticking out as if electrified by overflowing intelligence was introduced reverently to younger generations. His opinions were quoted with approval, his vision was admired, and his fame flared once more.

The mysteries that occupied Einstein throughout his life can be summed up in three questions: What is space? What is time? What is gravity? He did not answer them, of course, but he did discover relationships among them that had never been suspected, and he provided mathematical formulations that are radical and beautiful and, as far as we know, correct. His theory of gravity, worked out between 1910 and 1916, remains his greatest monument and one of the finest feats of pure reasoning in the history of natural philosophy.

Gravity, like space, is ubiquitous and, like time, cannot be turned off. Electricity, another familiar force, can be switched off; magnetism can be shielded; even the strong force that holds atomic nuclei together can be counteracted by antimatter; but gravity passes through all materials, affects all matter equally, and has no opposing force, no shield, no antigravity. Only God can turn it on and off, and He is proud of this prowess. "Can you bind the cluster of the Pleiades?" He rhetorically asks Job, who replies humbly: "I have spoken of great things which I have not understood, things too wonderful for me to know." Because it is always there, and because we cannot affect it, we are rarely conscious of gravity. Yet it dominates life. What triumph when the newborn infant first lifts her wobbly head to peer around—the first victory over gravity. From that moment on, life is a constant battle. We win decisively when we first stand up, learn to ride a bicycle, climb a mountain, scramble up a rope, hit a home run, erect a wall, build a dam, hang a painting, lift a dumbell, clear a hurdle, hoist a flag, or pull ourselves aboard a departing bus. Gravity, on the other hand, wins every time a pin drops, a plane crashes, a tower topples, an avalanche strikes, and a baby rolls off the bed.

More significant than these major encounters are the never-ending skirmishes that wear us down. Each day begins with a confrontation. We must rise from bed by lifting our bodies against the pull of gravity. Sometimes this little conflict escalates into a battle and ends in defeat. Gravity has captured another prisoner. At other times, proud of our early success, we taunt gravity by challenging it to a duel of pushups, knee-bends, or chinups. The outcome is inevitable: In the end gravity

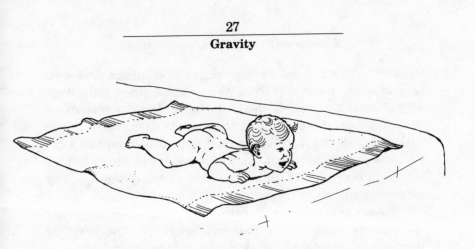

always wins. We spend the rest of the day climbing stairs, rising from chairs, lifting food to our mouths (and occasionally dropping it), moving pots or books around (and occasionally dropping them); in short, either lowering things that are up or raising things that are down. Meanwhile, the heart is pumping blood against gravity and the muscles are guying bones against collapse. The battle ends only when our bodies are finally abandoned to gravity in the grave.

The world is shaped by gravity and the operations of nature depend on it. After gathering the materials of the earth into a ball, it holds them together. Opposing the titanic convulsions of the young planet, gravity formed the mountains. It propels the rivers and streams, pulls the rain from the clouds and flattens the surface of the sea. It gives direction to the trunks of trees and the stems of flowers. Because of gravity, the lower parts of animals differ from their upper parts. Gravity acts as a restraining, organizing, direction-giving principle in nature. Inexorably it draws form out of chaos. It determines the shapes of stars and galaxies, the orbits of planets, and the expansion of the universe. It binds the cluster of the Pleiades and keeps our feet firmly planted on the earth.

The organizing role of gravity in the scheme of nature was engrained in Greek philosophy. Aristotle taught that the natural motion of heavy things is down toward the center of the earth. This is a most reassuring state of affairs. The mystery of gravity has dissolved. Why does a stone fall? Well, why shouldn't it? replies the Philosopher. Down is its natural tendency and it just follows that innate inclination. Much rather you should ask: Why does it come to rest on the floor? And the answer is that its natural tendency is checked there by the intervention of an artifice in the form of the opposing force of the floor.

To say that a thing is natural removes it from further speculation. Natural means normal, healthy, ordinary, rather than anomalous, pathological, in need of analysis and interpretation. The word "naturally" serves to end conversations, not to start them.

For almost two thousand years Aristotle's answer satisfied most philosophers. It was in the seventeenth century that Isaac Newton made gravity into something extraordinary, something to be aware of, something in need of explanation. To him, as it was earlier to Galileo, the natural state of an apple detached from its tree is to be at rest. Only under the influence of this special effect called gravity does the apple abandon its natural state and begin to fall. To us, who are earthbound, it seems strange that motionlessness should be called natural. An astronaut in mid-flight would find this idea more plausible because he is used to the sight of a hammer calmly remaining in place after he has released it.

Newton, by formalizing and generalizing the concept of gravity, deepened its mystery. He showed that gravity is not only a property of the earth but resides as well in the moon, the sun, and the planets. In fact, all material objects in the universe attract each other. The manner in which they do follows certain simple mathematical laws, which Newton ingeniously unraveled. But where does this force come from? What makes things attract each other?

The technical name for Newton's description is action-at-a-distance. It means that two objects exert a pull on each other without the need for contact or for an intervening medium. It is most strange. To influence another person we must use touch, or sound carried through the air, or we can send a letter, or at least let light reflect from our bodies to reveal us; but the earth influences us, pulls on us, without any such mediation. It pulls on the moon over a distance of thousands of miles. Action-at-a-distance is an idea far from common daily experience.

Newton had no illusions about his discovery. He knew that it explained nothing, even though it described the motions of the universe with exquisite precision. He believed that only God knows why, and was content that in His wisdom He had permitted a glimpse of His great system by revealing the how. Newton felt that gravity by action-at-a-distance is a mathematical construct that does not have to appeal to human intuition.

Lesser people do not share Newton's lofty detachment. Many have sought—and many are still seeking—models that make sense, that set the mind at ease when the awful question haunts them: Why do things attract each other? The history of those who rode forth to tilt against that particular windmill is long and colorful. It is marked by frustration, obsession, insanity, ignorance, and delusion.

Consider Cadwallader Colden (1688–1776). Born in Scotland, he became a botanist in the American colonies and eventually lieutenant governor of New York. (His competence was recognized by no less a man than Linnaeus, who called him *Summus Perfectus!*) In later life, without much understanding of the Newtonian doctrine that swept the world in the eighteenth century, he began to think about the problem of gravity. His aim was to answer the why that had eluded Newton. In 1745 Colden published a pamphlet in New York entitled "An Explication of the First Causes of Action in Matter," to which an English editor appended the tantalizing phrase "and of the Cause of Gravity." Colden's model was immediately rejected by those who understood physics, but it is remarkable in that it is reinvented with monotonous regularity to this very day. It explains gravity by pressure of particles of ether, which are assumed to fill the universe. Since the earth and the moon partially shield each other from the impact of those particles that come from far away, the pressure is smaller on the sides of the moon and earth that face each other. The two bodies are therefore pushed together. Cadwallader Colden died in the hope of achieving immortality, not through the little flower that Linnaeus named after him, but as the discoverer of the cause of gravity. His hope was not fulfilled.

Fortunately, most people are not possessed by wonder about gravity. Gullibility is a human foible. "What I tell you three times is true," said the Bellman a-hunting the Snark. Newton's explanation of gravity has been repeated so often and so authoritatively that we all believe it and indeed feel that its truth has become part of our intuition. William D. MacMillan, professor of astronomy at the University of Chicago, was moved to put it this way on the occasion of a debate on relativity at Indiana University in 1926, as he somewhat

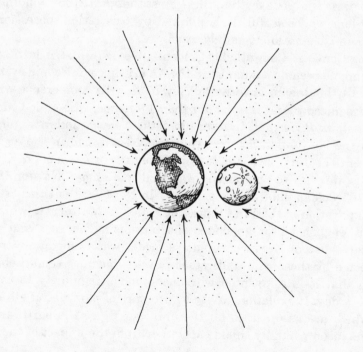

belatedly opposed Einstein's new theory: "The mechanics of Newton, like the geometry of Euclid, was based upon our normal intuitions and it is, therefore, intelligible in the normal sense of the word, just as Euclid is intelligible." Just as the Bellman said. But considered without prejudice, the notion of action-at-a-distance is disturbingly unsatisfactory. If it were intuitively obvious, it might have been invented sooner than two millennia after Euclid!

Sloth is another human foible. More than sixty years have passed since Einstein submitted a better description for Newton's action-at-a-distance. Because it is difficult to understand, we rarely hear about it. The old words are so much easier to repeat; General Relativity can be left to the experts. To be sure, many popularizers, including Einstein himself, have tried to explain it, but the Newtonian view still overwhelmingly predominates.

Mathematics, the language of Einstein's theory, is difficult for most people, even if they have no trouble with words. To symbolize the mathematics, analogies are made to familiar circumstances, but the images are of necessity imperfect. The most famous prediction of general relativity, and one that is easily amenable to analogy, concerns the deflection of starlight due to the curvature of space. Normally starlight reaches the earth in a straight line. Einstein's theory predicted that a ray of starlight that grazes the sun should be bent a little bit, giving the earthbound observer the illusion that the star's

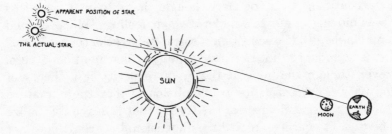

position has shifted. Because the sun is so brilliant, stars that are almost behind it—so that their rays graze it on their way to the earth—cannot normally be seen at all. The only opportunity comes when the sun is blocked out by a total solar eclipse. Astronomers looked for the effect as soon as possible after Einstein's prediction and confirmed it. The bending occurs because in the vicinity of the sun, space is curved.

Einstein himself, in his book written with Leopold Infeld, *The Evolution of Physics,* made an analogy to explain what is meant by curved space: "Imagine an idealized American town consisting of parallel streets with parallel avenues running perpendicular to them. The distance between the streets and also between the avenues is always the same. With these assumptions fulfilled, the blocks are of exactly the same size. In this way I can easily characterize the position of my block." This image represents ordinary or Euclidean space. Cars follow straight lines defined by streets and avenues. Imagine now that some subterranean upheaval causes a hill to bulge up in the middle of town, taking streets, avenues, and houses with it. The space represented by the grid of streets is now curved or non-Euclidean. A car would still follow the streets and avenues, but if it happened to be on a road that grazes the hill, its path would be bent a bit along a curve near the hill. In the same way, the sun causes curvature of space and a deflection of starlight.

The deflection of starlight and the curvature of space were the talk of the world in September 1919, when the experimental verification of Einstein's prediction was announced. One contemporary gushed later: "A wave of amazement swept over the continent. . . . The mere thought that a living Copernicus was moving in our midst elevated our feelings. Whoever paid him homage had a sensation of soaring above Space and Time." It is significant that the word *gravity* is not mentioned here, even though the general theory of relativity is a theory of gravity. The reason for this omission is that the curvature of space is one element of the theory that is easy to visualize because it concerns real three-dimensional space. It does not

Gravity

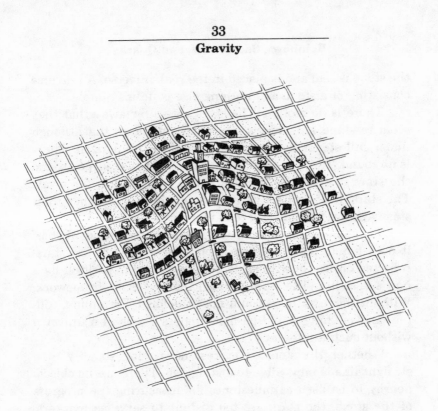

touch upon the true cause of gravity. To come to gravity, it is necessary to dig deeper and to invent other analogies, images that are increasingly blurred. The following word-picture is offered with some trepidation, in realization of its inadequacy.

Consider a stone in outer space. The size of a fist, hard and cold, it drifts in space. (Of course, its appearance makes no difference whatever, but to increase its palpability, imagine it smooth and shiny, a polished piece of fine marble, veins delicately etched in the pale matrix.) The earth, the moon, the sun, the stars and galaxies are far away, so far that their gravitational forces on the stone are too weak to be registered by even the finest instruments. Only the light from the distant stars, puncturing the translucent blackness, provides a link between the stone and the rest of the universe. The stars form a patterned background, like a vast cage, for the stone. This stellar cage is necessary. It cannot be imagined away, because the stone is real and is placed in the real universe. A universe consisting of a stone and nothing else is unimaginable.

There is no motion. The stars are so far away that they seem to stand still, like a ship in the distance that, though under full steam, seems to be at rest on the horizon. There is no sound. The vast clouds of gas and dust that surge around the stars and fill the spiral arms of galaxies are far away. The stars don't twinkle because no air breaks or bends their steady beams of light. There is no change.

The stone is very still. Needing neither support nor anchor, it does not tremble, roll, or pitch. The images of the stars, reflected on its polished surface, do not vary in position by a hair's breadth. (The stars provide the necessary framework. Without them, stillness could not be defined. Trembling, rolling, and pitching would be meaningless words in a universe without such a referent.)

Whether the stone is at rest or moving steadily in a straight line is impossible to distinguish. There are no objects nearby to be used as milestones for measuring the progress of the stone; the stars are too distant to serve as markers. On this point, Newton made a mistake. He would have imagined the solitary stone truly at rest within his own mental framework, which he called "Absolute Space." This edifice "in its own nature, without relation to anything external remains always similar and immovable," like a gigantic imaginary scaffold. The position of the stone would be measured along the fixed axes of Absolute Space. Newton justified his conception in the *Principia:*

> I do not define time, space, place and motion as being well known to all. Only I must observe that the common people conceive those quantities under no other notions but from the relation they bear to sensible objects. And thence arise certain prejudices, for the removing of which it will be convenient to distinguish them into absolute and relative, true and apparent, mathematical and common.

Thus he introduced the idea of a fixed space that needs no objects in it to provide reality. For over two centuries, Newton's lofty construction predominated, until Einstein tore down the scaffold and reintroduced those prejudices that his predecessor had struggled to remove. What for Newton was a false prejudice was for Einstein the truth. Motion can truly be thought of only in relation to "sensible objects"; without them it becomes meaningless. To illustrate this commonplace, Einstein, in the beginning of his first scientific paper on relativity, takes us to a homely railway station where a conductor is timing the arrival of a train by comparing the position of its engine with the position of the hands of his watch. Motion, to Einstein, is common, apparent, and relative, rather than mathematical, true, and absolute.

Consider the stone, then, in the bleak stillness of outer space. Nothing happens, nothing changes, nothing moves. To end the monotony, add something to the image. Close by, say ten thousand miles away, let the earth appear, materialized by the power of thought. Round, smooth, with white wispy veils of cloud over a bluish mottled surface, cool, silent, familiar and inviting, the space age vision of our home. In relation to this new neighbor, the true motion of the stone can be ascertained. The center of the earth provides a benchmark that now fixes the position of the stone, ten thousand miles away and motionless.

But the motionlessness is only momentary. Imperceptibly at first, then gathering speed, the stone begins to move toward the earth. More precisely, it falls toward the center of the earth. Gravity is at work. The general theory of relativity provides a picture of what is happening.

In Einstein's theory, gravity is related to another concept, hitherto unmentioned and apparently different in nature: the idea of time. That time must be considered here has rarely been said more emphatically than by Hermann Minkowski, who in 1908 ushered in the age of relativity in the ebullient preamble to his lecture on space and time: "The views of space and time which I wish to lay before you have sprung from the soil of experimental physics, and therein lies their strength. They are radical. Henceforth space by itself, and time by itself, are doomed to fade away into mere shadows, and only a kind of union of the two will preserve an independent reality." The union is called space-time and takes the place of Newton's Absolute Space as the stage for physical phenomena, including the fall of the stone in the void. Space-time is much further removed from our everyday intuition than is Absolute Space or even curved space. We are reduced to imperfect images.

Consider once again the stone without the earth. Nothing seems to be changing, but silently in the background there is now a gentle unfolding: Time is elapsing. Unlike space, which reaches up and down, right and left, forward and back, time flows relentlessly in one direction. To measure its flow, let a watch appear on the stone. The word "time" will be given meaning by the reading of the watch chained to the stone.

The complete union of space and time is unimaginable. The best we can do is to imagine time as a fictitious sort of space. The stone's history can then be thought of in borrowed words: The stone is moving through time. This phrase, almost trite from overuse in science fiction, requires amplification. For the sake of concreteness, a graph can be constructed of space and time. Position (in one dimension) is measured along one axis and time along the other. If a point on the graph represents a real object, like the stone in outer space, then

it will move along the graph as time progresses. Successive moments and positions are represented by successive points on the graph. Thus, by translating time into a position along an axis, as is done every day in the graphs on the financial pages of the newspaper, motion in time can be translated into ordinary motion in space.

The flow of time is represented by a flow of an imaginary medium: "Time is a sort of river of passing events, and strong is its current," wrote Marcus Aurelius. In the bleakness of the void, the stone (carrying its watch) floats in a vast, silent current of clear and subtle liquid that pervades every pore of the universe and bears forward everything within it. The current is time. Its motion cannot be stopped, its depth cannot be detected—because it is not real. Unlike a real river in real space, this current exists in four-dimensional space-time. The stone, like a stick of wood on the water, has no motion of its own, but drifts wherever it is carried by the imaginary river of time.

Finally, in our imagination, the earth appears again, ten thousand miles away. On the river of time, both stone and earth are carried along. At first they seem to travel in parallel lines, keeping their distance. But soon it becomes apparent that the streamlines are bent gently toward the earth. The current, its flow modified by the presence of the huge mass of the earth, carries the stone closer and closer to its neighbor. The bending of the streamline is barely perceptible at a great distance, but close to earth it becomes more pronounced. At the earth's surface, the flow is wildly distorted from its original direction.

This is an image of curved space-time, which is the crux of Einstein's theory of gravity. It is different and more subtle than the idea of curved space. The stone, when released, does not find itself in the mysterious grip of the earth, acting at a distance of ten thousand miles. Instead it abandons itself to the soft embrace of the river of time, which envelops it and quietly carries it along.

Thus we return to the harmonious Greek conception of

falling as a natural motion. In Newtonian physics, the earth does violence to the stone's natural inclination, which is to remain where it started. A force is needed to overcome its inertia. In Einstein's universe, the stone does what is most natural: It drifts along the curves of space-time toward the earth. With respect to the imaginary water of the river of time, it simply remains where it started.

Gravity, instead of pulling directly on objects far away, is mediated by space-time. The earth affects the streamlines, and the streamlines in turn guide the stone. Cause and effect are proximate: Each point affects only the surrounding points; they in turn pass the effect along the stream.

Einstein, who wrote equations rather than words, coaxed from them a number of definite experimental predictions. The most compelling one provided his motivation from the beginning. Galileo had observed that different masses fall at the same rate. (Air resistance modifies this fact a little, but let us pretend that there is no air and concentrate instead on gravity itself.) Since the pull of gravity is obviously stronger on greater masses (i.e., they are heavier), this observation is difficult to understand. Why shouldn't heavier things fall faster, just as Aristotle taught? Newton's explanation was that nature has devised a cunning conspiracy: Although a heavy stone experiences a stronger pull of gravity than a light one, it has just precisely so much more inertia, it is just precisely so much harder to budge, that the two stones end up falling at equal rates. This theory explains the facts, but it is contrived to give the right answer. How much simpler, in comparison, is curved space-time. Place into the stream of time a second stone, 10 times heavier than the first, and right next to it. The two will drift on and down toward the earth at precisely the same rate, because the current carries both together. A feather will do the same thing. And so will a piano or a grain of sand. The proposition that all objects fall at the same rate fits so effortlessly into the context of the curvature of space-time that the incomprehensible becomes obvious and the abstruse compelling.

Another prediction of the theory, more difficult to put into words, was Einstein's explanation of a tiny irregularity in the motion of the planet Mercury, which had plagued astronomers since the middle of the nineteenth century. Imagine the planet traveling around the sun in an ellipse, while both glide down the stream of time. Because the streamlines are curved by the mass of the sun, the path is not a perfect ellipse. The deviation is slight because Mercury always remains far from the solar surface where the curvature is strong.

The deflection of starlight, the equality of rates of fall, the motion of Mercury, and other successes of Einstein's theory impossible to duplicate in Newton's, have forever relegated action-at-a-distance to the shelf of historical curiosities. But there is no final answer. What is missing in the new thinking is an explanation of why a mass, such as the earth's or the sun's, distorts the streamlines. Once we accept the fact that they are curved, the motion of any body is easily understood, but why do masses curve space-time in the first place? Einstein provided an answer, but he was not very happy with it. In his equations, the curvature of space-time appears on the left and the mass of the earth or sun, which gives rise to it, on the right. He once likened the left-hand side to a marble palace, and the right-hand side to a wooden building attached to it. He left the task of reconstructing the annex to future generations.

Curved space and curved space-time are difficult ideas—

but they are better than action-at-a-distance. We could offer Albert Einstein no greater tribute than to move away from the rigid, cold, mystical view of Newton and toward his gentler and more homely way of thinking. Try it. Think of a stone in your hand. Imagine letting go. Now picture it carried silently, swiftly, along the river of time, which happens to have a little bend in it right here, directed toward the floor. The stone follows the streamline of space-time just as a twig follows a stream. What could be more natural?

The Rainbow

John Keats wrote in 1820:

Do not all charms fly
At the mere touch of cold philosophy?
There was an awful rainbow once in heaven:
We know her woof, her texture; she is given
In the dull catalogue of common things.
Philosophy will clip an angel's wings,
Conquer all mysteries by rule and line,
Empty the haunted air, the gnomed mine—
Unweave a rainbow.

The word *awful* sounds odd to our modern ear, but it is the only true word in these lines. Taken literally it recaptures its original rich meaning of solemnly impressive, sublimely majestic, commanding profound respect and reverential awe.

The rainbow's size, the intricate harmony of its perfect shape and colors, its very evanescence and consequent rarity combine to command universal admiration and wonder.

In classical mythology the feeling inspired by the rainbow was awe. When he described the rainbow as female, even though his metaphor is a ribbon of cloth, Keats was probably thinking of Iris, the Greek deity of the rainbow. The colorful part of our eyes, the metal iridium, and the iridescence of mother-of-pearl are named after her. Sometimes she is the personification of the rainbow itself, at other times she merely uses the arch as a bridge through the skies. In any case, she is a messenger, a comely young woman with wings and a herald's staff, flitting through the *Iliad* and later the *Aeneid* on little errands for the gods. Her appearance portends tumult and mischief for humans. Iris, the rainbow, was for the Greeks an omen of war and storms. She strikes terror in people's hearts, but at the same time she carries an air of mystery. She is the daughter of Thaumas, the god of wonder and marvel. Through him, Iris is related to magic, to miracle-working, to thaumaturgy.

A different sort of awe, closer to reverence than to terror, is associated with the rainbow in the Bible. In Genesis, God says to Noah: "This is the sign of the covenant which I establish between myself and you and every living creature with you, to endless generations: My bow I set in the cloud, sign of the covenant between myself and earth." Later in the Old Testament, the rainbow represents more than a covenant—it becomes an attribute of God. In Ezekiel we read, "Like a rainbow in the clouds on a rainy day was the sight of that encircling radiance, it was like the appearance of the glory of the Lord." Ecclesiasticus, in the Apocrypha, describes the rainbow among the wonders of creation, "a bow bent by the hands of the Most High." And in the New Testament, in the Revelation of John, God's throne is surrounded by a "rainbow, bright as an emerald."

Keats' past tense in "there was an awful rainbow once" is premature and will remain so. The charm and wonder of a rainbow will never cease. The feelings of the poet, the painter, the photographer, the meteorologist, the psychologist, and the physicist looking at the rainbow may differ, but an element of awe will always be there. The woof and texture of the rainbow were not completely understood in 1820; they are not entirely known today. Much has been learned, to be sure, but science is an infinite regression—behind each answer lurks a question, and behind that, another. The history of the physics of the rainbow, like all of science, is a continual unfolding. We know something of the rainbow's nature, but we will never unravel it all.

Understanding the rainbow, even in a limited way, means reducing it to causes that can be studied in the laboratory and captured mathematically. In that sense it is true that theories of the rainbow are catalogues of common things, paragraphs in textbooks, equations in technical journals. But whether they are dull or charming depends on the beholder. Richard Feynman, a marvelously free-spirited theoretical physicist, points out that Jupiter is known to be nothing but a great swirling mass of methane and ammonia, two common, even obnoxious, substances. Does this knowledge rule out Jupiter as a fit object for the contemplation of people with souls? Or, on the contrary, have the *Mariner* photographs of the planet, which resemble nothing so much as the paintings of Georgia O'Keeffe and which were made by computer scientists and engineers, added to the world's store of beauty and wonder?

The cold philosophy maligned by Keats is Natural Philosophy, now known as science. The rule and line with which it conquered the mysteries of nature in the past can serve no longer. Euclidean geometry was the preeminent branch of mathematics for physicists until the end of the nineteenth century, but today it is inadequate. Relativity and quantum mechanics have bent the rule and dispersed the line. Even when they were intact, rule and line did not conquer all mysteries; but in their present condition they pose more problems than they solve. Nevertheless, physicists continue on the quest, begun by philosophers two thousand years ago, to unweave the rainbow. They answer "No!" to Keats' question, and remain undeterred by his slanderous charge of angelic mutilation. They are sustained, rather, by feelings expressed by Wordsworth, eighteen years earlier:

> *My heart leaps up when I behold*
> *a rainbow in the sky:*
> *So it was when life began;*
> *So it is now I am a man;*
> *So be it when I shall grow old.*
> * Or let me die!*

The characteristics that make the rainbow awful are its rarity, its sudden appearance and unheralded disappearance, its enormous size, and its dazzling splendor. Like a huge spectre it comes and goes without warning. When sun and rain are both in the sky together, the rainbow *may* appear. When, after a rain shower, the sun breaks out of the clouds, quickly look the other way. The rainbow appears in the direction opposite the sun. It is perfectly circular, but unless observed from an airplane, part of the circle is hidden under the horizon. The sun, the head of the observer, and the center of the circle form a straight line that tilts down into the ground. The bow often reaches nearly halfway up to the zenith while the ends of the rainbow lie at an angle of almost ninety degrees to each other. The arch itself, a vivid band of color against the rainy sky, is more than four times as wide as the full moon. Its most brilliant color is red, always found at the uppermost edge. Orange, yellow, green, blue, and violet follow in descending order of brightness, delicately blending into each other and not always clearly distinct. (Isaac Newton suggested the interposition of indigo as a separate color between blue and violet. Later, in the nineteenth century, physicists amused themselves by counting thousands of hues in the continuum of colors.) Sometimes a second rainbow is seen higher in the sky, always with reversed order of colors. Very good photo-

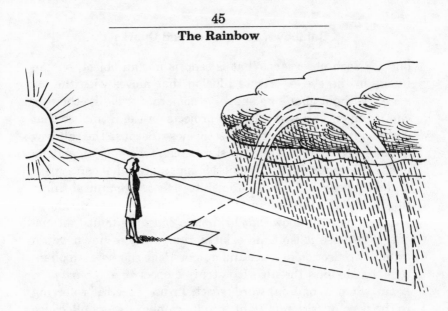

graphs show that the space between the two rainbows is distinctly darker than the rest of the sky. This region is called "Alexander's dark band," after its discoverer, Alexander of Aphrodisias.

Not all these subleties are familiar today, especially to city dwellers whose vision of the sky resembles a dirty blanket more than the crystalline background for celestial phenomena that it was for the Greeks. But even the simplest facts, thoughtfully examined, lead to startling conclusions. Consider this argument: The rainbow is always round; therefore it is not material. If the rainbow were a tangible object in the sky, like a painted wooden arch among the clouds, it would look different. From the front its appearance would be just what we are accustomed to, but at an angle it would be oval, like a McDonald's arch, and from the side the opening would be no wider than that of a hairpin. In fact, however, the rainbow never appears foreshortened into an elliptical shape. It is always perfectly semicircular, as though it were seen by each observer straight on. Each viewer sees exactly the same thing. This means that there isn't one rainbow: There are many, one for each observer. What is seen is not an object, but an image in the eye, a private illusion that moves with the observer without changing shape. A camera works like the eye, capturing on film what the eye projects on the retina, so that a photograph is no proof of objectiveness. Because the rainbow isn't really an object in the sky, it can't be caught and its ends are forever elusive. Because it isn't material, it can appear and disappear with such astonishing speed. A material thing, like a cloud, takes some time to cover the huge distances spanned by a rainbow, but an image can appear instantaneously. Because it isn't material, a rainbow does not have an ordinary reflection in lakes and mirrors. The rainbow is a phantom that haunts the air. (It is truly a spectre, a *spectrum* in Latin, whence come our words spectrum and spectral, referring to the decomposition of light into its rainbow colors.) Because it isn't material, its perfect symmetry is forever unspoiled.

The circular symmetry of the rainbow appealed to the Greek philosophers, for whom perfect circles governed astronomy and geometry. Since the rainbow is patently not an astronomical phenomenon, its circular shape must have a geometric origin. "God ever geometrizes" was the motto of the Platonic Academy. Aristotle's theory of the rainbow as a sort of reflection of sunlight by a cloud is unclear and fanciful, but his demonstration of the shape is correct and has survived without change. It is based on the observation that only a circle guarantees that the geometrical relation between the sun, the observer, and any point on the rainbow is exactly the same for *every* point on it.

Aristotle's argument is characteristic of appeals to symmetry that often furnish partial answers of powerful persuasiveness in contexts of obscurity and ignorance. A swollen ankle can be detected by comparison to its counterpart, even without any knowledge of anatomy or medicine. A men's room can usually be found in a strange building, once the women's has been spotted, by an appeal to symmetry. Symmetry allowed Archimedes to argue that equal weights balance a scale at equal distances from the fulcrum, even before the general case of unequal weights was understood. Similarly, it is possible for Aristotle to be right about the shape of the rainbow while at the same time quite wrong about its cause.

The explanation of the cause of the rainbow had to await the bold philosopher who snatched the spectre out of the heavens and brought it down to earth. Galileo examined heavenly motion by means of a rolling marble in his study; Newton applied celestial gravity to an apple in his garden; Franklin reduced lightning in the sky to sparks in his parlor. All three had motives exactly contrary to Plato's when he remarked that astronomy compels the soul to look upward and leads us from this world to another. They dared the heavens, and became immortal by bringing science down to earth, but the man who put the rainbow in the laboratory is almost forgotten now, because he lived three centuries too early. His name was Theodoric of Freiberg and his time the second half of the thirteenth century.

To put Theodoric's achievement into proper perspective, one must recall that the thirteenth century was the age of scholasticism. The greatest of the scholastics, Thomas Aquinas, died in 1274 when Theodoric was a student. Aristotle and the Bible were the two fundamental authorities of Western learning. Philosophical disputation was the preferred method of inquiry; its language sounds queer to us now. Much of it was concerned with *quiddity,* a medieval term referring to the real nature or essence of things. In fact Theodoric, like many other scholars, wrote a treatise entitled *On Quiddity.* Today the term has become pejorative. It means a captious nicety in argument, a quirk, a quibble. But behind the quiddities lay an earnest, sometimes desperate, struggle to get at the truth about heaven and earth. Nor was disputation the only available means to that end. Roger Bacon, a contemporary of Theodoric's, struggled valiantly against authority and tried to establish the roles of mathematics and experimentation in the pursuit of truth. It was Bacon who realized that each observer sees a different rainbow, and it was he who first measured the angular diameter correctly. But his wild spirit so angered authorities that his battles were rewarded by years of imprisonment. A handful of scholars besides Bacon dissented from the accepted doctrine on physics, but the majority followed Aristotle and his exegetes.

Theodoric was no rebel. He was an independent and wide-ranging thinker, and he engaged in his share of controversy, but unlike Bacon he stayed and advanced within the establishment. A German Dominican monk who taught at a convent in the mountain city of Freiberg near Dresden and studied in Paris, he later occupied high administrative offices, including that of provincial of Germany and German representative at General Chapters of his order. Theodoric wrote more than thirty books on logic, theology, metaphysics, psychology, and physics, of which over half are extant. Among them is *De Iride et Radialibus Impressionibus* (*On the Rainbow and the Impressions Caused by Rays*), possibly the greatest contribution of the medieval age to physical science.

Theodoric's discovery consists of two parts, both buttressed by meticulous, painstaking experimental and theoretical investigation. First is the idea that the rainbow is produced not by an entire cloud but by individual raindrops. This is the golden key that unlocks the secret of the rainbow. The resulting theory is completely radical. It is "atomic" in the sense of atomism as the ancient doctrine that all natural phenomena can be reduced to the actions of tiny irreducible agents. It ascribes macroscopic phenomena to the behavior of microscopic constituents. As long as investigations of the rainbow were under the authority of Aristotle, who had considered clouds and complicated cloud formations as the ultimate causes, they were doomed to failure as surely as theories of matter before atoms. (Of course the advantage of raindrops over atoms is that they are visible, but they are almost as elusive. When caught, raindrops lose their character, and in the sky they are just as invisible as atoms.) Theodoric's rainbow theory of 1304 is a vivid demonstration of the power of atomism.

In the spring of 1909, long after the explanation of the

The Rainbow

rainbow had become commonplace, the advantage of droplets over clouds was rediscovered in a completely different context. Robert A. Millikan and his students in Chicago were trying to measure the average electric charge on a drop of water by watching the motion of a cloud under the influence of an electric field in a cloud chamber. The method was crude and unsatisfactory. One day they substituted oil for water and to their astonished delight found their field of view filled with reflections from individual droplets dancing about in the chamber. (The fact that the droplets sparkled in all the colors of the rainbow added to the enjoyment, but had nothing to do with the business at hand.) This chance discovery led to the famous Millikan oil-drop experiment, the proof of the atomic theory of electricity, the measurement of the charge of the electron, and incidentally the Nobel Prize.

Millikan was able to arrest the motion of his droplets by adjusting the electric field; Theodoric did not have this power. Gravity pulls raindrops to the ground, relentlessly and without pause. This causes a problem, because the rainbow, rather than falling, stays majestically in place. To this objection Theodoric replied: "Although these drops, which are like watery spherulets, continually move downward in their fall, the fact that some drops succeed others in the same places causes this impression to appear to sight as being in the same place." A thoughtful objection effortlessly dismissed!

The Aristotelian proof of the circularity of the rainbow is incorporated in Theodoric's theory in sharpened form. It can now be said that the shape of the rainbow is completely determined by the geometry of sun, raindrop, and eye. As long as those three have the same relationship to each other, the appearance of the drop will be the same. Since all raindrops on a circle placed opposite the sun and centered on a line from the sun through the viewer's head form identical triangles with the sun and the eye, they all appear to have the same color.

The second part of Theodoric's contribution is an analogy. Because the raindrops in a cloud are too far away for experi-

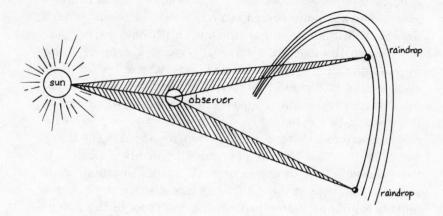

mentation, it is necessary to study an object that behaves *like* a raindrop, say a ball of crystal or a spherical bowl of water in the laboratory. The bowl is huge compared to the droplet, but it is a perfect analog. Glass and water spheres had been considered earlier in connection with the rainbow, but always as representing the cloud rather than its constituents. The mistake led Aristotle into a tangle of incredible assumptions about spherical clouds. Once the correct analogy was made, Theodoric plucked the rainbow from the sky and put it in his study. "The mode of this type of radiation," he wrote, "can be verified by experiment, if one uses a translucent crystalline stone, which is called a beryl, or any clear spherical drop that is so situated with respect to the sun and the viewer, namely with the viewer located between the sun and such a drop located off to one side of the straight line that goes from the sun to the viewer."

Today spheres, or at least cylinders of water, are easier to come by than in the Middle Ages. A brandy snifter full of water makes a good raindrop, but a glass with straight edges is easier to use because it produces, instead of a spot, a line of color. A flashlight lying horizontally, or a candle, can repre-

sent the sun. Darken the room, turn your back squarely to
the flashlight, hold the glass in your left hand at about 45
degrees from the forward direction, and look for the rainbow.
At first you see all sorts of reflections of the flashlight, but
no rainbow. That is as it should be. Reflection from the convex
front surface of the glass does not involve colors. For this reason the older theories, including Aristotle's, which involved
ordinary reflection, were wrong. But with a little patience, a
little lifting, turning, and tilting of the glass, suddenly you
find a splash of brilliant red near the right-hand edge of the
glass. That is the rainbow! It is caused by reflection of light
from the rear, concave surface of the glass. To produce it,
the light has to enter the water, bounce off the inside back
surface of the glass, pass through the water again, emerge
from the front surface, and travel to your eye. Whenever it
enters or leaves the water, the light beam bends, or refracts,
a little. The refraction on entering is toward the center of
the glass, on leaving toward the flashlight. By covering the
front surface of the glass with your free hand it is easy to
convince yourself that the light enters the glass near its left-

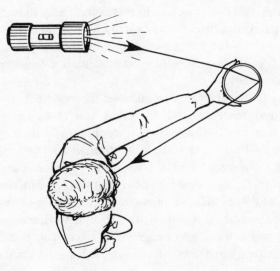

hand edge and leaves near the right. If you move the glass very slightly to one side, the red disappears. If you move it to the other side, other colors replace red, and then the image splits in two and fades away. The split image is a superfluous complication. What matters is that each color corresponds to a unique angle of view. If instead of moving the glass, you surrounded yourself with a ring of tiny glasses side by side, you would see red in one, orange in the next, then yellow, green, blue, and violet in order. You would see a rainbow in your room.

Rainbows are seen in diamonds, in drops of dew clinging to spider webs, in sprays from garden hoses, surfs and waterfalls, in ice crystals on trees. One can even blow water into the air with a puckered mouth and find a rainbow. The sequence of colors is always the same, the underlying physics always similar.

Theodoric, after explaining the primary rainbow, went on to explain the secondary bow with reversed colors, and the dark band of Alexander. The former is the result of two internal reflections. The light enters the raindrop, suffers refraction, bounces off the back surface, passes through the water, bounces once more, and undergoes a second refraction as it leaves. The dark band is, in retrospect, easy to understand. Raindrops cannot produce or destroy light. The light they concentrate in the bright primary and secondary bows must be missing from somewhere else: it was stolen from the space between the bows.

For all his spectacular success, Theodoric had to leave unanswered many questions about the rainbow. Principal among them are the quantitative problem of the angular size of the bow—that is, the angle subtended by the arch at the eye of the observer—and the question of color. Can the size be calculated from other known physical quantities? What causes the colors? Under modern circumstances, Theodoric's successors would have taken up the quest where he left off, building on his work and progressing beyond it. But Theodoric's misfortune was that he lived long before the printing press was invented. His books, meticulously copied and illustrated with diagrams resembling those of physics texts of today, quietly gathered dust in monastic libraries while the world revolved and the Middle Ages waned. The rainbow was discussed occasionally during the succeeding years, without much insight, until it turned up again at the very dawn of modern science, when Theodoric's discoveries had to be repeated.

The mechanism of the rainbow was described, in words and diagrams like those of Theodoric, by René Descartes in the seventeenth century. Descartes was able to calculate the size of the rainbow from the known optical properties of water, and it is customary today to give him credit for the first explanation of the whole phenomenon. The colors were explained a little later by Isaac Newton, and the supernumerary arcs, faint colored lines that occasionally appear at the inner edge, played an important role in the establishment of the wave theory of light by Thomas Young around 1800. More accurate and qualitatively different descriptions of the rainbow continue to be advanced. With the use of the computer it has become possible to explain subtle variations, such as the effects of droplet size and water temperature. In 1975 a sum of more than fifteen hundred complicated terms of a mathematical formula was used to describe the distribution in the sky of just one color of the rainbow. The reams of computer printout from this calculation undoubtedly qualify as a "dull catalogue." But the numbers cannot diminish the splendor of the rainbow, nor can they lessen the curiosity of the scientists who will continue to tease at nature's most brilliant braid without ever unweaving it.

Sky Colors

Why is the sky blue? Why is the sunset red? Why are the clouds white?

Curiosity, like hunger, is a powerful and necessary instinct. Without a sense of wonder, the mind would starve. When curiosity is allowed to flourish, it develops into the passion for knowledge and understanding that lies at the root of science. Not only the wording but also the motivation of a child is exactly the same as that of a professor of physics when they ask: "Why is the sky blue?" Both are simply curious.

Not all cultures encourage curiosity, especially of the intellectual variety, but in our civilization it is fundamental. If there is a credo of Western philosophy it is the famous maxim of Socrates, speaking to the court that condemned him to death: "The unexamined life is not worth living." Socrates wasn't sure that his listeners would understand or accept this proposition. Indeed, it is a radical idea, almost shocking in its peremptoriness. Other cultures, and powerful forces in our own, proclaim the opposite, that life should be taken as it comes, that contemplation is superior to examination, that the blue sky and the red sunset should be accepted and enjoyed without question. Socrates was thinking specifically of the spiritual life that must be examined, but to the extent that the external world is also part of life, his imperative includes a justification for science. To the scientist the unexamined world is not worth living *in*. The true scientist has to investigate, just as Socrates had to philosophize.

Most people don't feel a craving for scientific research, but many find in it a source of pleasure. Out of sheer curiosity they want to understand the phenomena around them. An indicator of this universal interest is the popularity of nature books, especially of those known as field guides. There are field guides on birds, snakes, mammals, beetles, butterflies, fish, shells, trees, flowers, mushrooms, fossils, rocks, crystals, clouds, and stars, among other assorted denizens of the out-of-doors. All are based on the premise that knowledge about the naming of things, and about their surroundings and their family relationships, enhances the enjoyment of them. Field guides are instruments of the pleasure of pure knowledge. They carry the legacy of Socrates.

Probably the best-known field guide, Roger Tory Peterson's *Guide to the Birds,* is found on the windowsills of homes that feature gardens without and curiosity within. It serves to identify wild birds for the edification of the old and the instruction of the young. Its users feel that their minds are broadened, and their senses sharpened, if they can identify the titmouse, the kinglet, and the towhee in a flock on the lawn and distinguish their calls in the woods. There is no practical use for this ability except to satisfy curiosity and to appease the hunger for knowledge.

Most field guides cover topics in natural history, which refers primarily to the biological sciences, but some physical sciences, such as geology and astronomy, are also represented. The student of these subjects for whom the field guides are written, is called a naturalist. In the seventeenth and eighteenth centuries, however, a naturalist or natural philosopher was specifically what we now call a physicist. That older usage suggests the question: Why is there no field guide to physics?

The answer cannot lie in a dearth of physical phenomena, because everything around us, living and inert, is continually engaged in physical processes. Gravity affects the animate as well as the inanimate, and the laws of motion govern the behavior of plants and animals as surely as that of boulders and stars. We are bathed in an infinity of colors of light, enveloped by sounds of myriad pitches, nourished by countless forms of energy, and driven to dissolution by the stern laws of thermodynamics. There is literally nothing that is unaffected by physics. Why then is there no field guide to a subject of such pervasive significance? If zoology, botany, geology, paleontology, chemistry, and astronomy inspire field guides, why doesn't physics?

The basic reason may be that field guides are at bottom taxonomic, while physics has moved beyond classification. In every scientific endeavor taxonomy must be the first order of business. The confusing welter of the things to be studied must first be put into some kind of systematic order. This is true of shells as well as stars, of spiders as well as snowflakes. The details of the scheme of classification are usually not terribly important, and they change as new insights are gained, but the initial step of sorting and naming is indispensable for a young science.

During the second half of the eighteenth century, and throughout the nineteenth, biology was dominated by the Linnaean system of classification of plants and animals. Field biologists roamed the world collecting specimens by the hundreds of thousands and sent them to European theoreticians for incorporation into the new framework. Without this systematic survey of familial relations, the theory of evolution, the basis of all biology, would not have been discovered. In astronomy the classification of stellar spectra by Annie Jump Cannon in the first quarter of this century led to the modern understanding of the structure and evolution of stars. In chemistry it was the taxonomy of the elements, in the form of Mendeleev's Periodic Table, that ushered in the modern era.

But physics has matured beyond taxonomy. As a science ages, the emphasis on classification wanes. As the underlying structures are understood and captured by simple mathematical models, long tables are reduced to brief formulas. Thus the compilations of elevations of cannons for different ranges were eventually summarized by the laws of projectile motion. The lists of angles of bending of light at the interface between glass and air have been replaced by Snell's law of refraction. The ancient catalogues of positions of planets have been reduced to a few numbers that specify elliptical orbits. Characteristically, taxonomy still plays an important role at the cutting edge of physics, where the science is young and evolving. The classification of hundreds of properties of elementary particles today plays the role of the Linnaean system in early biology. For the last twenty years the fraternity of high-energy physicists has used as badge of recognition a well-thumbed little pamphlet entitled *Particle Properties Data Booklet,* which is designed to fit in a pocket. It serves as field guide, but unlike most others of its genre it is written for professionals rather than amateurs, its setting is the accelerator laboratory rather than the out-of-doors, and it describes phenomena that cannot be discovered with the unaided eye. Particle physicists hope fervently that soon they will be able to deduce the hundreds of entries in the booklet from a simple mathematical model of matter, and that their field guide will become obsolete.

With the disappearance of taxonomy as the focus of a science comes a shift of the questions from What? to Why? While the amateur field biologist tries to match a flower to a description in a book, and at best to find discrepancies and surprises, the amateur physicist is challenged, after identifying a phenomenon, to relate it to other, more familiar observations or, better yet, to deduce its operation from fundamental laws. The question "What is the name of that little red bird?" is very different from "Why is the sunset red?" The distinction entails disparate arrangements of field guides for the two disciplines. Taxonomy automatically furnishes a scheme for a book that aims to sort out the phenomena. Textbooks on such deductive sciences as modern physics and Greek geometry, on the other hand, begin with a minimal set of laws and continue with derivations of the consequences in ever-increasing complexity. A priori it is impossible to know where in such a scheme the reasons for the colors of sunset, clouds, and sky make their appearance. They might occur at the ends of long chains of complicated argument, or they might be straightforward consequences of some simple law. They might be related, or they might be totally independent. Their surface similarities might hide profound differences. Or, what is most common, they might not be fully understood even today. A field guide to physics, organized in a traditional way, seems to present insurmountable structural difficulties.

And yet a sort of field guide to physics does exist. Written by Jearl Walker, editor of the "Amateur Scientist" section of the *Scientific American,* and entitled *The Flying Circus of Physics,* this book is obviously a labor of love and is destined to become as loyal a companion to the observer of the physical aspects of nature as Peterson's guide is to the watcher of birds.

Why is the sky blue? Why is the sunset red? Why are the clouds white? These and six hundred other questions, almost all starting with the word *why,* are listed. They range from the commonplace to the highly technical, from the well understood to the completely unknown. Why does the moon seem enlarged when it is near the horizon? Why does chalk squeal? Why do rivers meander? How does an airplane gain lift? Why do stars twinkle? Why do oil slicks on the street display colors? Why are golf balls dimpled? Why do Rice Krispies go snap, crackle, and pop? What causes the banging of steam radiators? . . .

Each question is elaborated in a paragraph, and some are illustrated by whimsical drawings. Instructions are given for observing the phenomena, simple experiments are suggested, hints are dropped as warnings against false explanations. The

questions are divided into chapters corresponding to traditional divisions of physics: Acoustics, Mechanics, Heat, Fluids,
Optics, Electromagnetism. Within each chapter, problems are
grouped according to their underlying physical principles, but
because the phenomena are so diverse and unexpected, it is
best to use the book for browsing or to consult the index.

The most ingenious feature of the work is the explanations. Because many are controversial, and most require elabo-
rate backgrounds, they are all left out. Instead, the reader is
referred to the mammoth bibliography of 1,632 items. Some
questions refer to a dozen or more sources, others to a single
one, and a few to none because there are none. This method
of answering—or rather, *not* answering—encourages speculation and original thought, followed by genuine library research
instead of a reach for the *Encyclopaedia Brittanica*. *The Flying
Circus of Physics* was conceived as a stimulant, rather than
an authority. "I am not so interested in how many [questions]
you can answer," writes the author in the preface, "as I am
in getting you to worry over them."

But alas, the noble purpose was defeated. The writer bowed
to the pressure for instant gratification, and two years after
its first publication a second edition appeared under the title
The Flying Circus of Physics WITH ANSWERS. And there they
are, a whole new section printed on blue paper for easy access.
The answers are sketchy, to be sure, and the reader is requested
to turn to them only as a last resort, but the damage is done.
Some of the charm, novelty, and mystery has left the book
along with the grammatical sense of the title. "Let's try to
find out" has been replaced by "Here's your answer, now leave
me alone." Nevertheless, as a field guide to the physical phenomena of the world around us, *The Flying Circus* remains
invaluable.

So what are the reasons for the red, white, and blue of
sunset, clouds, and sky? Let's try to find out. The most important step in explaining celestial phenomena, whether they are
mechanical like the motion of the moon, electrical like lightning, or optical like the rainbow, is to bring them into the

laboratory where they are amenable to experimentation. In the case of the colors of the sky, this is easily done. All that is required is to observe the appearance of a beam of light, say from a flashlight, filtered through a pitcher or fishbowl full of water into which a little milk has been stirred. By itself, the light bulb looks yellow, like the sun. Seen head-on through the milky water, it appears distinctly reddened. From the side, the cloudy water looks blue. By rotating the flashlight around the pitcher, while keeping the eye in the same position, a continuous change of color from blue to red and back again can be seen in the water. Different amounts of milk allow the display to be studied with different intensities.

The key to the understanding of many of the divers beautiful colors of the sky, and their reflections in lakes and oceans, is revealed in this simple experiment. None of the ingredients

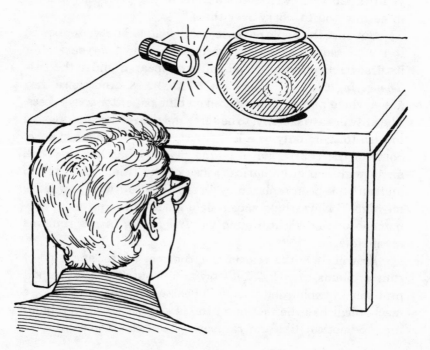

displays red or blue colors: Water is clear and translucent, milk is pure white, and the light is yellow. Nevertheless, red and blue are hidden in the apparatus. A prism would reveal that all the colors of the rainbow are mixed together in the light from the sun or from a flashlight bulb. What happens in the water is a process known as "scattering." Some of the light that enters the bowl is transmitted straight through, while another portion is scattered by the minute particles of milk and bounced out to the side, or even back in the direction from which it came. Evidently the amount of scattering depends on the color: The water appears blue because milk droplets are much better at scattering blue than other colors. What passes through is yellow light robbed of its blue component. Since blue and red appear at opposite ends of the rainbow spectrum, the electric light without blue is unbalanced and appears redder than the original mixture.

Experiments with various substances have shown that the color that is deflected most efficiently depends on the size of the particles responsible for the scattering. For very small particles, like globules of milk, blue light is scattered preferentially. For larger particles, the effect becomes less pronounced, until a size is reached that scatters all colors equally well.

These observations lead to the assumption that the sky is blue not because it shines in that color, nor because there is a blue backdrop behind it, but because the atmosphere scatters the blue component of the sun's light. When the sun is high, its light traverses a thin slab of atmosphere overhead. This layer is not thick enough to deplete sunlight of much of its blue. In the evening, however, the light from the setting sun slices obliquely through the atmosphere and follows a much longer path in air. More blue is scattered out, leaving only red. Thus, arguing by analogy to the kitchen experiment, both the blue sky and the red sunset can be explained in terms of one fundamental process: the scattering of light from small particles.

Since the world is immersed in air as well as in sunlight, the interactions between air and light manifest themselves

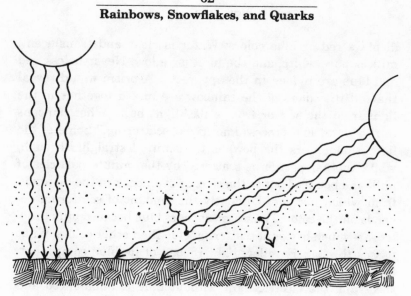

in a great variety of ways besides clear skies and sunsets. Painters who try to reproduce the infinite universe of hues and shadows on canvas are better equipped than anyone else to appreciate the subtlety of color and light in the open air. It was Leonardo da Vinci who, combining science and art into a single exuberant celebration of nature, first sorted out and described some of the basic observations.

One of Leonardo's lasting contributions to the theory of painting was distinguishing among three kinds of perspective. The first, called linear perspective, deals with the diminution in size of objects as they recede from the eye. This subject had been thoroughly treated even in Leonardo's day, and is familiar. The second and third, called color and aerial perspective respectively, were new. Color perspective refers to the change in color of objects as they recede: "The greater the depth of the transparent layer which lies between the eye and the object, the more will the color of the object be modified by the color of the intervening transparent layer." Aerial perspective was the term used by Leonardo to describe the increasing haziness of distant objects: "The impact of the appearance and of the substance of things diminishes with every successive degree of remoteness; that is, the farther the object is from the eye, the less will its appearance be able to penetrate the (layer of) air (which lies between it and the eye)."

Innumerable specific observations and sketches illustrate Leonardo's statements of these two laws. In modern terms, they deal with the phenomena of scattering and of absorption of light by air. Absorption simply reduces the amount of light so that the whole appears dimmer and less distinct the farther away the object is. More fascinating and delicate is the change in color with distance. At first guess it might seem that everything in the distance ought to appear redder than at close range, just as the sun looks redder in the evening than at noon. Indeed, for lights and fires, and for bright objects like snow-capped mountains, this is true, but for dark areas the effect is reversed. The eye, looking at a distant forest, detects only a small amount of light reflected by the green trees, but it also records the sunlight scattered by the intervening air. The scattered light, like the sky, is blue. Thus dark objects look increasingly blue and more uniform in color the further away they are.

The blue of distant mountains is a lovely sight. In hilly terrain each succeeding crest is fainter and bluer. More than a dozen individual peaks throughout the world called "Blue Mountain," the "Blue Ridge" of the Appalachian chain, and the "Blue Mountains" of Oregon, Maine, Jamaica, and Australia bear witness to the phenomenon, which is accentuated by the presence of vapors exhaled by trees. Roman wall paintings show blue mountains in the distance, but the effect was apparently forgotten until the Renaissance. The backgrounds of the paintings of the Flemish masters of the fifteenth century are famous for their magnificent shades of blue. Use of color perspective culminated in the nineteenth century with William Turner, the master of water, air, and light, who refined it into an eloquent pictorial language. Poets have painted it in words:

> *What are those blue remembered hills,*
> *What spires, what farms are those?*

That is the land of lost content,
   I see it shining plain,
The happy highways where I went
   and cannot come again . . .

laments A. E. Housman. Blue, the color of life-sustaining air, and by reflection of water too, has a special appeal to the soul. According to John Ruskin, "Blue is everlastingly appointed by the Deity to be a source of delight."

A whole new world of air colors is revealed to the passenger of a high-flying jet. The fanciful shapes of white and gray clouds, the pale sky above, and the dark patches of land and distant water commingle to form a variegated symphony in blue. Turner once, against the strenuous advice of his friends, had himself lashed to the mast of a ship for several hours in order to observe at first hand the colors of a tempest at sea. What would he not have given for a window seat on a 747 from London to New York!

The explanation of the blue color of the sky as a consequence of scattering rests on an analogy, but many questions remain. What are the little particles that scatter the light? In the case of the home experiment, they are droplets of milk, but what causes the scattering in the sky? Could it be the air molecules themselves?

Hypotheses are promoted to the status of theory, and later enshrined as facts, if they yield predictions that can be verified. Qualitative predictions are useful, but to be really convincing they must be quantifiable. Natural philosophy turns into physics at the point where observations become measurements.

In the case of the color of air, there exists a wonderfully simple device for attaching numbers to the words. Consisting of nothing more than a cardboard mailing tube with small holes in the end-caps, it is called a nigrometer to signify its function of measuring the degree of blackness. If one looks through the tube at a distant perfectly black object, like an open window, the nigrometer reveals a distinctly blue spot. The color is that of the air between the observer and the win-

dow. If a little mirror is mounted at an angle of 45° at the
end of the tube, it is possible to look simultaneously forward
at the window and up at the sky. Now let the observer walk
toward or away from the window until the two images coincide
in brightness. Since the mirror reflects only one-twentieth of
the amount of light incident on it, the observer knows that
the amount of blue light scattered into the tube by the atmosphere is exactly 20 times greater than that bounced back
by the air column between eye and the window opening. The
distance to the window, multiplied by 20, should be the thickness of the total atmosphere if it were compressed to its density
at ground level. This "equivalent thickness of the atmosphere"
is well known from other, independent measurements. Here,
at last, is a prediction. If the two numbers are roughly equal,
then the assumption that the color of the atmosphere is due
to molecules of air is justified.

The nigrometer was invented, and the experiment performed, by the American physicist Robert Williams Wood in
1920. The reported numbers agree.

R. W. Wood was something of an enfant terrible in physics.
A brilliant experimentalist, he specialized in physical optics
and made important discoveries in that field. His lasting fame,
however, derives from his whimsical sense of humor and his

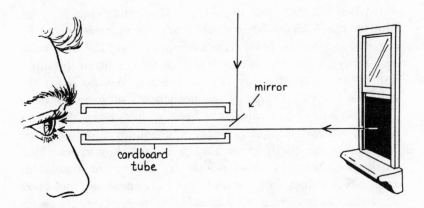

practical jokes that haunted laboratories around the world. His articles and books are now footnotes in the literature, but one work promises immortality to its author. It is nothing less than an interdisciplinary field guide, the only one available on the difficult subject of flornithology. Entitled *How to Tell the Birds from the Flowers*, its drawings point out the remarkable similarities in appearance of pairs of animals and plants. Entries range from "The Plover. The Clover" to "The Tern. The Turnip." The verse following each pair of pictures clarifies the differences for the beginner. For example, two cigar-shaped objects are distinguished thus:

The Parrot and the Carrot one may easily confound,
They're very much alike in looks and similar in sound,
We recognize the Parrot by his clear articulation,
For Carrots are unable to engage in conversation.

If there is a point to this charming silliness, it is the warning against the confusion of phenomena that exhibit superficial similarities but differ profoundly in nature. The opposite is often true in physics. The blue color of the sky and the red of the setting sun are as unlike as can be, but both can be traced to the same underlying cause. It is this unifying power of science that Wood parodied in his cheerful contribution to naturalism.

Wood's nigrometer does more than illustrate in ingeniously simple fashion the mechanism of atmospheric scattering. It changes our perception by proving that the blue color of the sky and of distant mountains really resides in the intervening air itself. The air is not invisible: We see it all the time. When we look up, we are not staring off to infinity. Instead, we are looking at a thin layer of blue fluid, the atmosphere, against the pitch-black background of outer space. Mountains are really green and gray, but they appear blue in the distance because we see the air in front of them. This realization makes our world at once more confined and more secure. Protected from the vast cold darkness of infinite space, we wander happily about in the earth's blue amniotic fluid.

Waves

A traveler who happened to be on the road from Utrecht in central Holland to the nearby town of Maarsen on the third of June, 1845, would have encountered a curious and unforgettable sight. Here, as in almost every other European country during that decade, the fresh new track of the first railroad cut like a ruler through the summer landscape. Alongside the rails little clumps of men, spaced at regular intervals, seemed to be preparing for some important business. Attired in gleaming top hats, flaring frock coats, and tight checkered trousers, some carried notebooks, others watches and a few, to the traveler's astonishment, trumpets and bugles. Suddenly the little scenes sprang to life. In the distance, trailing a long black plume, a train appeared, consisting only of an engine, a coal tender, and a flatcar. On the car, and even on the tiny open locomotive, men with notebooks, watches, and trumpets were already in action. Periodically a few isolated, brief notes blared out, sounded alternately by the musicians on the embankment and those who rushed by briskly at thirty miles an hour. After each set, the observers, stationary and moving alike, consulted briefly among themselves, argued busily, recorded something, and prepared for the next event. Once the whistle of the engine blew its strident note and everybody listened attentively. After the train had passed, the puzzled traveler, if he had been of a patient disposition, would have seen it stop in the distance, reverse, and return at a different speed—again accompanied by trumpets, bugles, craning of necks, strained listening, arguing, and scribbling.

What strange ritual was this? An inaugural ceremony, without crowds, without flags, without music save the solitary dissonant notes from the horns? Some arcane piece of engineering practice, without taciturn technicians, shouting foremen, and leathery workers, the only actors having the pale sedentary look of professional musicians? In fact, it was neither. The whole mysterious happening was a large-scale scientific experiment of the greatest significance to the future of physics and astronomy.

Conceived and conducted by Dutch meteorologist Christopherus Henricus Didericus Buijs Ballot, the experiment was designed to test the proposition that the pitch of a note depends on whether its source is stationary or moving with respect to the listener. Today we are familiar with the effect because the speeds of vehicles are greater and consequently the change more obvious. Who has not heard a train-whistle that seems to sing out a haunting wail even though the horn produces only a single note, or the curious downward modulation of the roar of a passing truck? In 1845 these sounds had not yet been heard. As tools for his experiment Buijs Ballot, perhaps because he was a meteorologist used to the out-of-doors and to phenomena as large as the sea and the mountains, picked a railway train rather than a sophisticated arrangement of more delicate apparatus in a laboratory. Indeed, his results were soon reproduced in conventional settings by experimental physicists, but for the purpose of popularization his dramatic demonstration was decidedly more effective.

Theory predicts that compared with a stationary source of sound, the pitch of an approaching source is higher and that of a receding one lower. If a trumpet on a moving flatcar emits a note, an observer standing near the track should hear a higher note before the train has passed and a lower note afterward. The same should occur if the trumpet is stationary but the observer moving. The shift in pitch is zero when the relative speed is zero, and increases with increasing speed. The experiment on the Utrecht–Maarsen line was designed to verify these predictions.

The report of the attempt, published in a contemporary journal, is refreshingly forthright. Helpful collaboration by the Minister of the Interior and the Director of the Rhine Railroad in securing free use of the train is gratefully acknowledged at the outset. Engine noise that obscured the first set of observations, loss of a notebook from the train, the inability of some observers to distinguish quarter notes, the failure of the engineer to keep a steady speed, an unfortunate tendency of the best observer to leave out important information in his records, the suspicion that several listeners mistook the train whistle for a trumpet, fear that some instruments changed pitch during the course of the experiment, the muddles in the lists of more than a hundred individual tones— all these frustrations are carefully reported. In the end, however, the conclusion is unequivocal: Motion changes the pitch of musical notes in just the predicted way.

Stranger than the experiment itself was the motivation. The real purpose of Buijs Ballot was the experimental corroboration of a theory that had appeared three years earlier in a paper written by Austrian physicist Christian Johann Doppler, "Concerning the colors of double stars and a few other luminous celestial objects." Double stars, like the two suns in the movie *Star Wars,* sometimes display distinctly different colors. A lovely example is the binary Albireo, which appears to the naked eye as a single white dot in the constellation of the Swan, and which can be resolved by binoculars into two stars in close proximity, one red and the other blue. Double stars revolve around each other, joined by their mutual gravitational attraction like dancers in a waltz, but since a single revolution may take days or even years to complete, the motion is not perceived at a glance. Doppler tried to find a causal relationship between the motion and the differences in colors of double stars. He reasoned that if the earth happens to be in the plane of the orbits of the two stars, then at any moment one star is approaching and the other receding. He then derived what has become known as the optical Doppler effect: A source of light looks bluer than normal if it is approaching the observer, redder if it is receding.

In addition to binaries, Doppler considered other stars with distinctive colors. Might they not also acquire their hues by the same principle if they happen to be moving toward the earth or away from it? Ironically, while Doppler's law is correct, his explanation of stellar colors is false. The effect that bears his name is real, but the colors of the stars, whether paired or single, have a different origin. The resolution of this apparent contradiction requires quantification of the problem. Approaching stars do indeed look more blue, but by such a minute amount that only sensitive instruments can measure the change. The real cause for the yellow, or red, or blue appearance of a star is related to its composition and temperature, not to its motion. The strong intrinsic color outweighs the tiny shift occasioned by the Doppler effect.

This paper about the color of stars prompted the experiment on the railroad. Doppler himself had pointed out the link between two phenomena as seemingly unrelated as the sound of trumpets and the light of stars: Both are described in terms of waves. While the wave nature of sound and light is far from obvious, waves themselves are familiar. We see them on the ocean, in amber fields of grain stirred by the wind, in flags in the breeze, and in the flames of candles trembling in a draft. Their harmonious regularity exerts a soothingly hypnotic effect on the receptive observer.

Waves are a curiously futile kind of motion, and therein lies their fascination. They move on relentlessly in perfect illustration of Heraclitus' maxim that nothing endures but

change—all is flux and reflux without beginning or end. Waves on the shore, for example, flow in, hour after hour, day after day, century after century, but water does not accumulate on the beach nor is the ocean depleted. In an ocean wave, the water molecules bob up and down and back and forth, returning over and over again to their starting positions. The ears of wheat and the fabric of a flag are not carried away by the wave motion, yet the waves themselves continue to surge inexorably forward. In contrast, the motion of a ball or a bullet has a beginning and an end; at the conclusion of the motion, an object has left one point and arrived at another.

The proposition that sound consists of waves is not difficult to accept because occasionally one can almost feel or see them. The lowest notes of an organ shake the body in a most palpable way. Some periodic disturbance, some vibration, must be broadcast from the organ pipe to the listener. Loud music rattles and shakes bodies, furniture, and walls without the accretion of any kind of material. Not only the recipients but also the sources of sound vibrate. A violin string, a tuning fork, and the diaphragm of a loudspeaker are visibly in motion when they emit sounds. They in turn shake the air, which transmits a periodic disturbance to the ear.

A medium, such as air or water, is required to carry sound;

in a vacuum all is still. During the eighteenth century, when vacuum pumps were popular toys in the drawing rooms of Europe and America, little bells were placed under the glass vessels of the pumps; their sound grew fainter and finally stopped as the air around them was withdrawn. One can speculate that the deafening roar that must accompany the prodigiously turbulent upheavals of boiling gases on the surface of the sun would be perceived on earth as a constant murmur that rose and fell with daylight if interplanetary space were filled with some kind of fluid instead of being almost totally void.

Sound, then, is a wave excited in the air in the same way that a tapping finger dipped in a sink full of water induces a wave on the water's surface. As long as the wagging persists, the wave continues. In the case of sound, the molecules near the source are shaken up, they bump into their neighbors, and so transmit a disturbance across a room until the molecules of air in the ear are reached and in turn bombard the eardrum. No stream of air accompanies the spreading of the sound wave—only a local vibration of the carrier medium.

Light, too, consists of waves, but the evidence for that proposition is less accessible than it is for sound. For centuries two ideas battled for supremacy: the particle theory, which posits the existence of subtle atoms of light, and the wave theory. A good argument for the former is the ability of light, in contrast to that of sound, to traverse a vacuum. Since there is no air in outer space, but light reaches the earth from distant stars, does it not seem reasonable to think of light as a beam of particles? If in fact it is an undulation, what is the medium that is undulating?

And yet light does behave like a wave. The best evidence for this side of the debate is the phenomenon of destructive interference. Waves have the peculiar property that they can cancel each other. If two identical ocean waves meet in such a way that the troughs of one lie exactly on the crests of the other, and vice versa, the net result at the location of the encounter is a flat ocean surface. To be sure, once they have passed through each other, the original waves are reconstituted, and whatever energy disappears at one spot crops up in another. Nevertheless, the fact remains: Two waves can create a calm region devoid of disturbance. Streams of particles can also pass through each other, but they cannot be mutually annihilative. The equation wave + wave = zero can, under certain conditions, be correct. The analogous equation, particles + particles = zero, is always false.

Destructive interference can be observed in all waves if you know where to look. For light, an easy experiment is as close as the tip of your nose. Find a bright source of light in the shape of a line—for example, an incandescent bulb with a single straight filament. A point source, such as a bright light bulb at some distance in the darkness of the night, works almost as well. Place the index and middle fingers of your right hand, almost touching each other, in front of your right eye and close to it. Steady those fingers by holding their tips firmly with your left hand. Now observe the light through the narrowest part of the gap between the two fingers. Vary the size of the gap from zero, when the fingers touch, to an opening as thick as a match.

The light will appear with alternating bright and dark

bands arranged symmetrically on either side of it, running parallel with the fingers. These fringes are neither figments of the imagination, nor images of your eyelashes, nor consequences of the construction of the eye. As the gap becomes narrower, they spread farther apart; as it opens up, they crowd together—until between well-separated fingers they disappear into the image of the bulb itself.

The bright ribbons are evidence of constructive interference: Crests of waves meet each other to create higher waves. The dark bands are crucial—they are caused by destructive interference of waves of light that come to the eye from different parts of the gap between the fingers, and they cannot be explained on the basis of a particle theory. If particles were responsible for the bright bands, they would fill in the dark areas as well and no striations would be observed. Thus a look between your fingers can demonstrate the invisible structure of light.

Interference suggests that light consists of waves, but the particle theory, first proposed by the ancient Greek atomistic philosophers, is by no means dead. Albert Einstein pointed out in 1905 that there are certain atomic experiments in which light behaves not like a wave but very much like a particle. In such cases it arrives in discrete bundles that carry identical fixed amounts of energy, which exert a push on the object they strike, and which occupy small spaces, more like raindrops than ripples on a lake.

This state of affairs sounds schizophrenic: Light can be proved to be undulatory and it can be proved to be particulate. Both sides of the venerable argument appear to be right. The way out of the dilemma is radical: Stretch the mind and accept both sides simultaneously. It is only the macroscopic world that displays sharp boundaries between waves and particles. In the microworld of atoms the distinction is blurred so that both descriptions must be applied to the same object. The situation is reminiscent of the discovery of the platypus. When explorers brought back tales of a mammal that lays eggs, the learned world scoffed: Reptiles are reptiles, mammals are mammals; one animal can't be both. But nature is full of things that no one dared to imagine. The platypus does exist. It has characteristics of both mammals and reptiles, and it transcends the categories that had been established before its discovery. Thus, too, light is described as both a particle and a wave.

The converse applies also: Just as light has particle properties, atoms and their constituent particles exhibit wavelike behavior. Experiments analogous to peering between fingers have been performed for electrons, with the result that they are observed to display interference just like waves. The history of physics in the twentieth century is to a large extent the experimental and theoretical elaboration of this duality between waves and particles. Phrases such as "wave mechanics," "wave equation," and "wave function" have replaced the Newtonian "equation of motion" and "particle mechanics" in the working vocabulary of atomic and nuclear physicists.

The image of waves has proved to be a powerful theme in physics. From the swells on open seas and ripples on lakes it reaches into the realms of sound and light where the evidence is necessarily indirect, and thence down to the subatomic world where it becomes a mathematical metaphor, a model, capable of making verifiable predictions but without the immediate sense appeal of waves in the world of everyday experience.

Doppler, taking advantage of the analogy between sound and light, suggested that one could be used in place of the other to test his predictions concerning moving sources. In order to understand the experiment, waves have to be quantified. Numbers must be attached to natural phenomena before the physicist can analyze them mathematically. The two numbers that characterize waves are speed and frequency. The speed of a wave, like that of a car, is simply the distance that each crest traverses in a given period of time. Frequency, on the other hand, is the number of crests of a wave that pass a stationary observer in that time period; it indicates how closely the crests follow each other. High frequency signifies tight spacing, low frequency a greater distance between successive maxima and minima. For sound waves, frequency determines pitch in an intuitively plausible way. Higher frequency means higher pitch. The whine of a circular saw ripping into a board rises in pitch when the speed of the blade, and consequently the frequency with which the teeth strike wood, increases. For light, frequency is associated with color in a less obvious way. The frequency of a wave of blue light is high, that of red light, low.

Radio and television signals consist of waves that resemble light, and travel with the same speed, but differ in frequency from station to station. AM frequencies are measured in thousands of cycles per second, or kilocycles, FM by millions of cycles per second, or megacycles. The abbreviations kc and Mc appear on older radio sets. The modern name for a cycle per second is "Hertz" in honor of the discoverer of radio waves—whence кHz and MHz. Television channels are also characterized by frequency, but the channel numbers are arbitrary and do not represent a physically significant quantity as they do in radio.

With the aid of the concept of frequency, the Doppler effect becomes a simple matter. Imagine a seashore pounded by a steady surf. If, on the average, six breakers hit the beach every minute, the frequency of ocean waves is six crests per minute. If a motorboat sets out from the shore and makes for the open sea, its bow will hit more than six waves per minute. A passenger in the boat would observe a greater frequency when moving toward the source (the open sea) than when onshore. If, on the contrary, the boat were to approach the land, moving more slowly than the waves but in the same direction, fewer than six crests per minute would sweep by the prow.

The Doppler effect expresses the fact that frequencies for an observer approaching a source are increased, and for a receding observer are decreased. For sound, this implies a change in pitch, for light a change in color, for radio reception a change in the tuning dial. The directions of the changes are toward higher pitches, bluer colors, and higher frequencies for approach; redder colors, lower pitches, and lower frequencies for separation. The same explanation applies to sound, light, radio, ocean, and particle waves, as well as to all the myriad other wave phenomena in the world.

Buijs Ballot verified the acoustic Doppler effect, but nature had made use of it long before the first train pulled out of Utrecht. Bats emit beeps of high-pitched sound that bounce off a target and return to the sender. The elapsed time before detection of the echo indicates the distance to the target. In addition, the bat's brain analyzes the pitch of the echo. If it is higher or lower than the original beep, the Doppler effect is in operation and the bat can derive the speed of the target relative to itself. Police officers do exactly the same thing, using

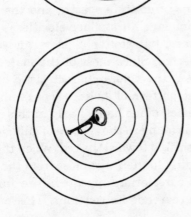

radio waves in place of acoustic signals to catch speeders. In both cases only the results are significant. An understanding of the Doppler effect is not required by either bats or cops.

Ironically, the grandest application of the Doppler effect is made in astronomy, but not at all in the way its discoverer had intended. The frequencies contained in starlight can be measured accurately by means of a prism that spreads an incident beam into a spectrum. If the spectrum included all colors of the rainbow, the Doppler effect would not have an appreciable effect because each color that was removed from its usual position by being shifted a little bit would be replaced by another color shifted into the vacant spot. Since there is invisible radiation beyond both ends of the visible spectrum, even the displacement of the ends would be undetectable. The Doppler-shifted spectrum would look exactly like the normal one. Fortunately for astronomy, the spectrum of starlight is far from uniform. Instead of emitting a continuum of rainbow colors with equal intensity, a few discrete colors are favored. When split by a prism, a beam of starlight exhibits individual, well-separated bright lines of color over a general faint background. Each line comes from a specific element in the hot surface of the star, and each one has a definite standard frequency that can be measured in the laboratory. If the star is in motion, every line is shifted with respect to its normal position by a measurable amount indicative of the star's speed. Nature, by means of discrete spectra and the Doppler effect, provides a speedometer for stellar velocities.

Not only stars, but stellar clusters and entire galaxies emit discrete spectra. Since most of the material in the universe is hydrogen, the colors characteristic of that element predominate. The pattern of its lines is as familiar to an astrophysicist as the back of his hand. Early in this century a striking regularity was found in the spectra of galaxies. All, without exception, were shifted toward the red. When the galaxies were sorted out according to their distances from the earth, an even more wonderful correlation appeared: The further a galaxy is from us, the greater is its red shift. This means, applying the Doppler effect, that all galaxies are rushing away at speeds that increase with increasing distance from the earth. We seem to be the pariahs of the universe.

A simple, almost trivial model makes sense of this frightening realization. A rubber band, 10 cm long, is fixed at one end. The other end is pulled so that the band stretches. Assuming that in one second it doubles its length, the speed of the free end is 10 cm per second. The speed of the halfway point, which moved from a point 5 cm to one 10 cm from the fixed point in one second, is only 5 cm per second. Each point of the band, in fact, moves with a speed proportional to its distance from the fixed end. If this one-dimensional image is generalized to a huge three-dimensional cake, in which raisins are embedded to represent galaxies, and if the whole cake is allowed to expand uniformly, a plausible model of the universe emerges. An observer sitting on a raisin notices that all the other raisins are receding with speeds that depend on their distance. Most intriguing is the fact that this same observation is made from the point of view of each raisin, regardless of its position in the cake.

The conclusion is one of the most presumptuous scientific pronouncements that mere mortals, sitting in cozy armchairs by the fire, have dared to make: The universe is expanding. Like the drawn-out lament of a distant train whistle, its pitch lowered by the speed of the receding engine, the faint glimmer of a remote galaxy, reddened by the Doppler shift, sends its farewell across the immensity of outer space. In the baffling field of cosmology, where questions about the size of the universe, the nature of its boundaries, the beginning and end of time, the curvature of space, and the infinity of stars strain the minds of professionals as much as those of amateurs, one fact stands out with unambiguous certainty: The universe, whatever may be its other attributes, is expanding.

From trumpet blasts on the Utrecht–Maarsen line out to the edge of the universe, from tempestuous billows on the Pacific to suburban radar traps and the shrill cries of bats, down to the elementary particles that compose the minuscule nuclei of atoms, waves obey identical laws. Identical mathematical equations governing their behavior render the analogies real rather than merely heuristically suggestive. The enormous scope of physics is one of the themes that distinguishes it from other fields of human endeavor. The same laws apply to objects that range in size from the diameter of a proton to the radius of the known cosmos, i.e. by a factor of 10^{40}, or one followed by forty zeroes. The same gravity that binds the universe operates between an atom and its neighbors. The swirl of milk in a cup of coffee echoes the eddies of gases in a galaxy. The collisions of stars, seen from afar, are indistinguishable from the collisions of billiard balls and of atomic nuclei.

In contrast, the scope of other disciplines is much more limited. Laws of nations bear little resemblance to rules of conduct for families, although the ratio of sizes in this case is a mere factor of a hundred million, or 10^8. The economy of a village in Australia has little in common with that of Japan. A biologist who specializes in elephants has little to say to a bacteriologist, even though the same underlying chemical and genetic processes link all organisms.

The vastness of the scope of physics is related to the essential role of mathematics as its language. Mathematics is a consistent and severely orderly system. By its means, physics finds or imposes consistent and orderly patterns on the chaotic world of experience. Where both similarities and differences exist, it highlights the former and minimizes the latter. By searching for the fundamental, physics captures the universal.

The universality of the laws of physics extends not only through space, but through time as well. It is a necessary article of faith that fundamental laws are the same everywhere in the universe, and that they have not changed since the beginning of time. Without these assumptions, or themata, there would be no cosmology. If Newton's law of gravity or the description of waves differed from place to place, or changed through history as human laws do, no valid inferences could be drawn about the universe, because whatever evidence there is can be interpreted only according to what is known here and now. In cosmology the questions of place and time are intimately related because of the long time required by light to travel to earth from the farthest galaxies. The faraway and the long ago merge into one, and it is assumed that physics there is as we know it here.

If the relation between motion and frequency, derived theoretically by Doppler and verified by Buijs Ballot, was valid ten billion years ago in a place that is now ten billion light-years away, then the red shift tells us that the universe is expanding. Putting things that way, the peculiar little scene near Utrecht is endowed with significance of cosmic proportions. Conversely, it is inspiring to realize that the patterns of the universe are accessible to human understanding and reveal themselves in circumstances as homely as a trumpet on a flatcar.

Whirlpools

And the waters prevailed exceedingly upon the earth; and all the high hills, that were under the whole heaven, were covered. Fifteen cubits upward did the waters prevail; and the mountains were covered. And all flesh died that moved upon the earth, both of fowl, and of cattle, and of beast, and of every creeping thing that creepeth upon the earth, and every man.

The destruction of the world is described in Genesis in a few scant verses. Only an active imagination fortified by a conscious effort to face the implied horror can penetrate beyond the terse lines to a fuller perception of the magnitude of the calamity. Recently such efforts have been made by writers and moviemakers who embellished the cryptic technical statistics concerning nuclear explosions with appalling details of the consequent devastation. Long before them, however, when the extinction of humanity was still the sole prerogative of divinity, the deluge and the day of doom fired the creative fantasies of sensitive artists who expressed their dread in painting, poetry, and music.

One of these, perhaps the most single-minded, was Leonardo da Vinci. Throughout his life he was fascinated by water. Its appearance, its manifold forms of movement, its uses in irrigation and navigation, its necessity for life, its destructive forces, its paths through the sky and over the ground, its colors, its interactions with air and with fire, its beauty, its terror— all these served Leonardo as objects for artistic and scientific study. A modern compilation of his writings on water comprises almost a thousand paragraphs and still remains incomplete. Leonardo observed water with the keen eyes of a painter, a physicist, and an engineer simultaneously. Late in life, at a time when thoughts turn to mortality and eternity, he gathered and brought to bear all that he had learned on one final problem: the depiction of the Deluge.

Nothing can take the place of an extended quotation of Leonardo's instructions for painting the Flood. They represent a gloss on the brief biblical passage and a powerful illustration of the consistency and unity of Leonardo's artistic and scientific visions.

> Let the dark, gloomy air be seen beaten by the rush of opposing winds wreathed in perpetual rain mingled with hail, and bearing hither and thither a vast network of the torn branches of trees mixed together with an infinite number of leaves. All around let there be seen ancient trees uprooted and torn in pieces by the fury of the winds. You should show how fragments of mountains, which have been already stripped bare by the rushing torrents, fall headlong into these very torrents and choke up the valleys, until the pent-up rivers rise in flood and cover the wide plains and their inhabitants. Again there might be seen huddled together on the tops of many of the mountains many different sorts of animals, terrified and subdued at

last to a state of tameness, in company with men and women who had fled there with their children. And the fields which were covered with water had their waves covered over in great part with tables, bedsteads, boats and various other kinds of rafts, improvised through necessity and fear of death, upon which were men and women with their children, massed together and uttering various cries and lamentations, dismayed by the fury of the winds which were causing the waters to roll over and over in mighty hurricane, bearing with them the bodies of the drowned; and there was no object that floated on the water but was covered with various different animals who had made truce and stood huddled together in terror, among them being wolves, snakes and creatures of every kind, fugitives from death. And all the waves that beat against their sides were striking them with repeated blows from the various bodies of the drowned, and the blows were killing those in whom life remained.

Some groups of men you might have seen with weapons in their hands defending the tiny footholds that remained to them from the lions and wolves and beasts of prey which sought safety there. . . .

Herds of animals, such as horses, oxen, goats, sheep were to be seen already hemmed in by the waters and left isolated upon the high peaks of the mountains, all huddling together, and those in the middle climbing to the top and treading on the others, and waging fierce battles with each other, and many of them dying from want of food.

And the birds had already begun to settle upon men and other animals, no longer finding any land left unsubmerged which was not covered with living creatures. Already had hunger, the minister of death, taken away their life from the greater number of the animals, when the dead bodies already becoming lighter began to rise from out of the bottom of the deep waters, and emerged to the surface among the contending waves; and there lay beating one against another, and as balls puffed up with wind rebound back from the spot where they strike, these fell back and lay upon the other dead bodies.

Following this piteous description of suffering, the technical details are laid out:

But the swollen waters should be coursing round the pool which confines them, and striking against various obstacles with whirling eddies, leaping up into the air in turbid foam, and then falling back and causing the water where they strike to be dashed up into the air; and the circling waves which recede from the point of contact are impelled by their impetus right across the course of the other circling waves which move in an opposite direction to them, and after striking against these they leap up into the air without becoming detached from their base.

And where the water issues forth from the said pool, the spent waves are seen spreading out towards the outlet; after which, falling or descending through the air, this water acquires weight and impetus; and then piercing the water where it strikes, it tears it apart and dives down in fury to reach its depth, and then recoiling, it springs back again towards the surface of the lake accompanied by the air which has been submerged with it, and this remains in the slimy foam mingled with the driftwood and other things lighter than the water, and around these again are formed the beginnings of the waves, which increase the more in circumference as they acquire more movement; and this movement makes them lower in proportion as they acquire a wider base, and therefore they become almost imperceptible as they die away. But if the waves rebound against various obstacles then they leap back and oppose the approach of the other waves, following the same law of development in their curve as they have already shown in their original movement.

The rain as it falls from the clouds is of the same colour

as these clouds, that is on its shaded side, unless, however, the rays of the sun should penetrate there, for if this were so the rain would appear less dark than the cloud. And if the great masses of the debris of huge mountains or of large buildings strike in their fall the mighty lakes of the waters, then a vast quantity of water will rebound in the air, and its course will be in an opposite direction to that of the substance which struck the water, that is to say the angle of reflection will be equal to the angle of incidence.

About five years after the passage was written, Leonardo died without having executed his plan. However, he left a large number of drawings and sketches, of which two illustrate his notes with particular vividness. The first is a technical drawing of water gushing out of a pipe into a hollow, in which it creates a whirlpool. The second is of the Deluge. In spite of the vast difference in scale of the two scenes, their similarity is obvious.

The whirlpool looks like the model for Leonardo's verbal description of how the Flood should be painted. Its most conspicuous features are spirals consisting of long gentle curves followed by tight whorls, a theme characteristic of all the water studies. The coils are clustered in separate strands, much like muscles and arteries in the anatomical drawings. In both cases the mind of the artist clarifies and separates the individual elements of what looks to the untrained eye like a wild, confusing mess of foam or flesh. Leonardo was perfectly aware of the fact that flesh, blood vessels, muscles, bones, sinews, and tendons are in fact organized in separate cords, while streaming water is not. The actual motion of a clear fluid is invisible, but, as Leonardo pointed out, it is made apparent by leaves, dirt, bubbles, and other floating objects. Throughout his work he draws streamlines as comprehensible visual shorthand for depicting flows.

The most surprising feature of the drawing of the whirlpool is its plasticity. The motion of the water is shown simultaneously in horizontal and vertical eddies. Leonardo's pen follows him beneath the surface where only the mind can penetrate. A whirlpool, captured by the eye or the camera, reveals only a superficial roiling. The drawing, however, illustrates the words: ". . . piercing the water when it strikes, it tears it apart and dives down in fury to reach its depth, and then recoiling, it springs back again towards the surface of the lake accompanied by the air which has been submerged with it." Rings of bubbles surround the little geysers where the spiraling coils of water, rebounding from the bottom, break through the surface. The drawing is more than a snapshot of a scene of wild motion; it is at the same time a theoretical treatise on fluid flow.

The drawing of the Deluge represents an enormous raincloud from which the waters pour in thick strings shaped exactly like the coils of the whirlpool. The masses of light and dark in the two pictures, and the detailed elements that make them up, resemble each other closely, but where the impression of the vortex is cool, analytic, and briskly effervescent, the Flood evokes a sense of threatening power and wild tumult.

The circumstances and scales are different, but the shapes of the currents are similar. Intuitively Leonardo had grasped a principle that dominates the modern study of fluid motion. In 1883 Osborne Reynolds, watching water and dye flowing through a glass tube, discovered the law of similarity. The flows of water through two tubes of different size, he found, resemble each other, provided that the flow through the thicker tube is slower than the other in the same proportion as the diameters. More generally, rapid flow of small dimensions is similar to languid flow of larger size. The similarity even extends to other types of fluids such as oil, paint, honey, gas, and air. The flow patterns of a sluggish, syrupy, viscous material are the same as those of a thin medium flowing more slowly.

Reynolds' surprising principle summarizes the conditions under which flows of different speeds, dimensions, and viscosity exhibit similar behavior. It allows us to see a spiral galaxy in a coffee cup and a tornado in the smoke of a cigarette. It enabled the Wright brothers to design their airplane based on data gathered in a wind tunnel no bigger than a breadbox. It captures mathematically the important realization that the analogies observed by artists and scientists among the most diverse phenomena of fluid motion are not accidental at all. They constitute, rather, a universal law of nature.

A rich variety of patterns is found even in simple circumstances, such as the motion of fluid past a circular cylinder placed perpendicular to the direction of flow. The arrangement is illustrated by the swirl of water around the cylindrical pier of a bridge over a river. Leonardo, with customary faithfulness, drew such eddies but, since he was an observer of nature rather than an experimentalist, he could not follow the changes in the pattern as the speed of water or the size of the obstruction varied. For him there was only one river, flowing at one speed. But the modern scientist, equipped with a laboratory flow tank and a camera, can study a series of snapshots taken at different speeds.

A wonderful collection of photographs of this kind was assembled in 1982 by Milton Van Dyke into *An Album of Fluid Motion*. The flow is rendered visible by light bouncing off impurities that follow the fluid. In the case of water they include milk, dye, aluminum dust, magnesium shavings, and air bubbles. Oily mist and cigarette smoke are used to follow the flow of gases. These devices replace the dirt and leaves of Leonardo, and, in connection with the camera, show directly what he imagined and then depicted by pen and pencil: Fluid follows streamlines that, while having no tangible reality, provide a fictitious stationary guiding grid for the flow. In this they resemble the lines that mark the routes of ships across the oceans.

When the water is flowing extremely slowly, the pattern of streamlines around the circular obstacle is completely symmetric fore and aft as well as right and left. The water approaches the obstacle, parts evenly to flow around its left and right sides, and converges again on the other side. The point of divergence is situated symmetrically to the point of conflu-

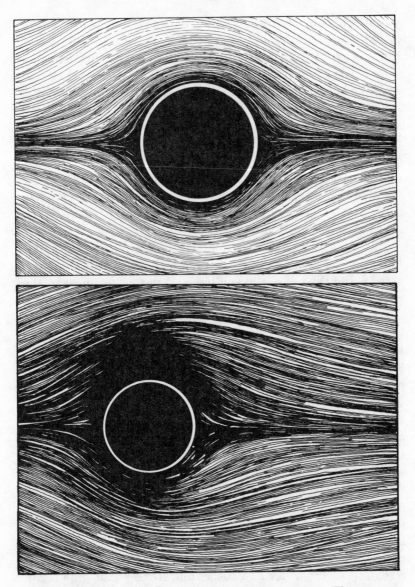

ence. At slightly higher speed the symmetry is lost. The streamlines still diverge upstream of the obstacle, but recombine further downstream. Thus the second photograph, in contrast to the first, reveals the direction of flow. At still higher speed a beautiful new phenomenon sets in. Immediately behind the cylinder two eddies form on either side of the center line, mirror images of each other. The water flows rapidly past the sides of the obstacle and then circles back toward the sheltered rear surface, where it is deflected once more out into the stream. Leonardo's drawings of these pairs of vortices correspond in fine detail to modern photographs. They are so accurate that a physicist, if he knew the scale of the drawings, could use Reynolds' law to deduce the speed of the water that flowed past Leonardo almost five hundred years ago. As the speed of the water is increased further, the whirlpools elongate until they reach far downstream of the obstacle, maintaining all the while their lenticular shape.

If the speed continues to increase, again something new happens. Up to this point, the eddies have been stationary, attached to the back of the cylinder. Eventually, however, the rush of water becomes so great that they are torn away and carried downstream. This happens alternately in perfectly regular periodicity: first the left-hand vortex grows, detaches, and leaves, then the right one takes its place until it too departs. Far downstream they dissolve into the steady flow. A snapshot of the pattern reveals a lovely sequence of whirlpools alternating clockwise and counterclockwise like a braid trailing from the cylinder. Known as a vortex street, it can be seen occasionally from bridges over swift streams. Its mathematical derivation has been accomplished only in this century.

At higher speeds, the vortices grow in size as they float downstream, so that the braid unravels and loses its streetlike appearance. Finally, a different phenomenon altogether: A speed is reached at which the steady flow is disturbed, first in the center of each moving vortex and then in the entire region downstream from the cylinder. Instead of following streamlines, the water now rushes and swirls about chaotically without apparent shape or pattern. Turbulence has set in.

As the velocity of the medium increases, its flow past the obstacle displays four distinct stages: smooth flow, stationary eddies, moving eddies, and turbulence. Of these, the first is

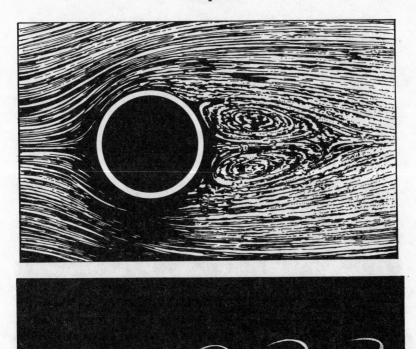

fairly uninteresting and the last almost incomprehensible, but the whirlpools of the middle stages have always exerted a powerful fascination on the human imagination.

Aristophanes, spoofing science in *The Clouds,* has Socrates explain how thunderstorms are caused not by Zeus, but by a natural vortex of air. The reference is to the ancient philosophy of Anaxagoras who taught that the cosmic process began when Mind, "the finest and purest of things," set up a whirlpool process that resulted in the separation of opposites. Much later,

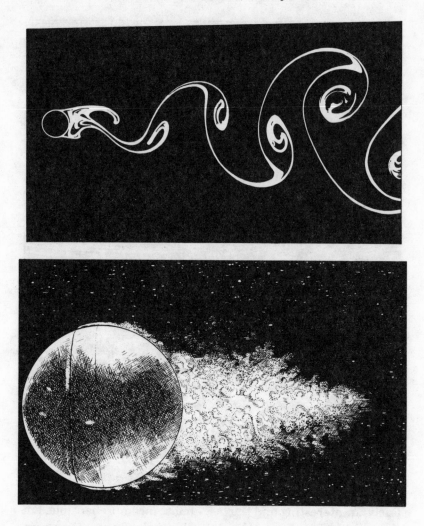

Descartes failed in an attempt to build a complete philosophy of nature based on vortices, but his hypothesis concerning the formation of the planetary system out of a whirlpool of primordial dust was probably close to the truth.

A marvelous fictional vortex is found in Edgar Allan Poe's "Descent into the Maelstrom," inspired by a real Norwegian fjord known for its treacherous whirlpools. The frightening description of the interior of the funnel was certainly not derived from observation, but Reynolds' law allows extrapolation from the whirlpool created by an ordinary bathtub drain.

Powerful natural vortices such as hurricanes and tornadoes are uncontrollable, but small eddies can be tamed. Since their suction exerts a drag on the obstruction that causes them, it is often useful to try to prevent vortex formation. One technique for doing this is streamlining. If a cylinder in a stream is replaced by a teardrop shape that fills only the space formerly taken up by the paired downstream vortices, the fluid flows past without eddying and with very little drag. Streamlining has been applied to all kinds of rapid vehicles. It accounts for the strange shields on top of the cabs of trucks, and for the bulbous noses that have begun to appear under the waterline of large ocean vessels. To a significant degree streamlining has shaped twentieth-century design.

Nature, too, makes use of streamlining to reduce fluid friction. Bowing to the dictates of the rules of fluid flow, evolution has seen to it that most fishes have similar basic shapes. Some mammals, such as dolphins, follow the same design, even though their inner workings and development bear no resemblance to those of the fish. Thus the laws of physics restrict the lavish freedom of possibilities with which biology is endowed and enforce common solutions to common problems faced by very different organisms.

More troublesome than stationary eddies are those that are shed into vortex streets. Since they leave alternately from right and left, the drag changes sides and can subject its generator to violent buffeting. The most dramatic instance of the destructive potential of this phenomenon was the collapse of the Tacoma Narrows suspension bridge in 1940; the bridge was set into oscillation by vortex shedding in a steady wind and rocked into self-destruction in less than an hour. That a steady wind could set a bridge in motion seems at first glance paradoxical. It is impossible, for example, to pump up a child on a swing by means of a stationary fan. But the mechanism of vortex shedding introduces a regular pulse into the smooth flow. If the bridge happens to resonate to this beat, it may begin to shake. Since the disaster at Tacoma Narrows, bridges have been stiffened to resist the onslaught of vortices.

Vortices are amenable to mathematical treatment, which can be verified in the laboratory. Turbulence, on the other hand, is much less tractable; even after decades of intense study, we are only beginning to understand it theoretically. The problem is so difficult that it prompted Sir Horace Lamb to remark in 1932: "I am an old man, and when I die and go to heaven, there are two matters on which I hope for enlightenment: one is quantum electrodynamics and the other is the turbulent motion of fluids. About the former I am rather optimistic." Indeed, we know today that the Lord could present to Sir Horace a definitive textbook on the quantum theory of electrodynamics, but we don't know what He would say about turbulence.

The sudden onset of turbulence is as familiar as cigarette smoke and water faucets. A cigarette left in an ashtray in a quiet room produces a stream of smoke with a characteristic pattern. For the first few inches above the glowing tip the flow is straight, smooth, and steady. Then it breaks up into wild, unpredictable eddies. Since turbulence in water was found to set in at relatively high speed, it is reasonable to assume that the hot smoke, as it rises into the air, moves faster and faster until it attains the speed at which turbulence sets in naturally. In kitchen faucets, on the other hand, turbulence is introduced artificially. A powerful jet of water from a tap feels hard and unpleasant. Since speeding it up is impractical, another technique, suggested by Reynolds' principle, is used. Instead of the speed, the dimension of the flow is changed. A wire mesh divides the single stream into many tiny ones, which become turbulent at low speeds and by their chaotic motion entrain bubbles of air that soften the stream.

Although turbulent flow is a homely occurrence, its theoretical description is at the frontier of knowledge among the unsolved problems of physics. It seems that new techniques of mathematics and statistics will be required to come to terms

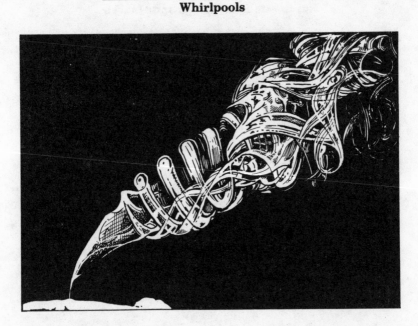

with its immense complexity. Among these, of course, are computers. As their powers grow, they can simulate the roiling of a turbulent stream of water with increasing faithfulness. However, there is a price to pay. Unless new general laws are discovered to describe the behavior of the water, simulation is not very helpful because computer codes ultimately become so huge and complex that they are just as incomprehensible as turbulence itself.

Recently some surprisingly simple mathematical equations have been discovered that also exhibit the transition from orderly behavior to chaos. Their solutions can be studied by means of digital computers. It is possible to construct electrical circuits in such a way that the current that flows through them obeys the original equations. When the switch is turned on, the measured signal exhibits the expected approach to chaos. This method represents a strange blurring of the distinction between the observer and the observed. Is the circuit a physical system that we observe and describe mathematically in the traditional way of theoretical physics? Or is the circuit an analog computer that merely serves as a tool for solving mathematical equations?

The latter point of view, if adopted, can be applied to a turbulent stream. A whirlpool becomes an analog computer that automatically solves the complicated equations of fluid flow. In order to read off the solutions we need only to observe the water and measure its flow. The study of turbulence, which began with Leonardo's clear eye and sure hand, returns to its starting point; the spiraling whirlpool becomes a metaphor for its own scientific explanation.

Lightning

In the middle of the eighteenth century there occurred in Russia a tragic accident of the kind seen when scientists, carried along by their own hubris, forget that nature commands forces far greater than any that art can produce. It was reported quickly throughout the learned world by letters and newspapers, serving as a reminder that nature's mysteries must be probed with humility and respect. The news saved the lives of many professional and amateur scientists, and possibly of countless others who had no idea of its true significance. The event was the death by lightning of physics professor Georg Wilhelm Richmann in St. Petersburg on the afternoon of July 26, 1753.

The history of the experimental study of lightning had begun three years earlier with Benjamin Franklin's plan for a method of demonstrating its electrical nature. The hypothesis itself, which is based on the fairly obvious similarity in appearance between a laboratory spark and a bolt of lightning, had occurred independently to many scientists as far back as Newton. It was the practical and bold American who suggested in a letter to the Royal Society of London that a man standing on a high tower, equipped with a long metal rod, might pull down some of the electricity from a cloud. Since there was no convenient tower in Philadelphia at the time, Franklin did not perform the experiment. Nor did the English physicists take up the suggestion. In France, however, all of Franklin's experiments were followed with avid enthusiasm, and Thomas Francois d'Alibard decided to try this one too. On May 10, 1752, at twenty minutes past two in the afternoon, his assistant succeeded in drawing long electrical sparks from a forty-foot iron rod erected in a garden in Marly-la-ville near Paris. The experimental proof that thunderclouds are electrified was justly praised at the time as the "greatest discovery that has been made in the whole compass of philosophy since Sir Isaac Newton."

Benjamin Franklin, in the meantime, thought of a substitute for a tower and in June of the same year, before the news of the French success reached him, flew his kite with similar results. In characteristic fashion he immediately put the discovery to practical use by publicizing the lightning rod, which he had in fact invented three years earlier on the basis of unproved speculation. By the end of 1752 there were lightning rods not only on Franklin's house, but on many public buildings and churches in the American colonies.

Contrary to the notion that technological innovation spreads more quickly today than it did in earlier times, the experiments with atmospheric electricity and the lightning rod were soon known throughout Europe and America. In Russia they were taken up by two physicists, a German and a Russian, who were the best of friends. They were Georg Richmann and polymath Mikhail Vasilievich Lomonosov, founder and eponym of the University of Moscow, professor of physics, astronomy, chemistry, and metallurgy, discoverer of the law of conservation of mass, convinced atomist, anticipator of the mechanical theory of heat, dramatist, poet, codifier of the Russian language, and national hero.

Lomonosov and Richmann had repeated Franklin's earlier experiments and described them in the *St. Petersburg News.* In 1753, a year after the historic experiment at Marly, they set up at their respective homes what they called "thunder machines"—lightning rods connected by wire and chains to a measuring device of Richmann's invention that consisted of a silken thread hanging from the upper end of a vertical iron rod marked as a ruler. When a thundercloud passed overhead the thread and the rod would receive similar electrical charges and repel each other. The thread, hanging loosely, would be pushed away from the rod at an angle. By measuring the angle the scientists hoped to quantify the strength of the electricity in the cloud or, in their words, the "degree of electrical force emitted by the cloud."

On the morning of the fateful day Richmann and Lomonosov were preparing a presentation on electricity for a planned public convocation of the Academy of Science. When a thunderstorm approached, Richmann hurried home, accompanied by

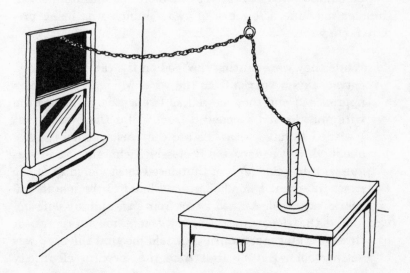

the academy's engraver, who was engaged in drawing the electrical phenomena. Thus it happened that a trained observer became an eyewitness who could later describe the event:

When the professor had looked at the electric indicator he judged that the thunder was still far off and believed that there was no immediate danger; however, when it came very close, there might be danger. Shortly after that the professor, who was standing a foot away from the iron rod, looked at the electric indicator again; just then a palish blue ball of fire, as big as a fist, came out of the rod without any contact whatsoever. It went right to the forehead of the professor, who in that instant fell back without uttering a sound onto a box standing behind him. At the very same moment followed a bang like the discharge of a small cannon, whereat the master of engraving fell to the ground and felt several blows on his back. It was later discovered that they came from the wire, which was torn to pieces and which left burned stripes on his caftan from shoulder to skirt.

Lomonosov was at that very moment examining his own thunder machine a few blocks away. Lunch was being prepared. He wrote:

While they were putting the food on the table, I obtained extraordinary sparks from the wire. My wife and others approached and they as well as I repeatedly touched the wire and the rod suspended from it, for the reason that I wished to have witnesses see the various colors of fire about which the departed Professor Richmann used to argue with me. Suddenly it thundered most violently at the exact time that I was holding my hand to the metal, and sparks crackled. All fled away from me, and my wife implored that I go away. Curiosity kept me there two or three minutes more, until they told me that the soup was getting cold. But by that time the force of electricity greatly subsided. I had sat at table only a few minutes when the man servant of the departed Richmann suddenly opened the door, all in tears and out of breath from fear. I thought that someone had beaten him as he was on his way to me, but he said, with difficulty, that the professor had been injured by thunder. Going to his home with the greatest possible speed my strength allowed, I arrived to see him lying lifeless. His poor widow and her mother were just as pale as he. The death which I so narrowly escaped and his pale corpse, the thought of our friendship, the weeping of his wife, his children and his household, affected me so deeply that I could say nothing and give no answer to the great number of people who had assembled as I looked at that person with whom I had sat in conference an hour ago and discussed our future convocation. The first blow from the suspended ruler with the thread fell on his head, where a cherry-red spot was visible on his forehead; but the electric force of the thunder had passed out of his feet into the floor boards. His feet and fingers were blue and his shoe torn but not burned through. We tried to restore the movement of the blood in him, since he was still warm; however, his head was injured and there was no further hope. And thus he verified, by a lamentable experiment, the fact that it is possible to draw off the electric force of thunder; this must be by directing it, however, onto an iron staff which should stand in an empty place where the thunder can strike as much as it wishes. Nonetheless, Mr. Richmann died a splendid death, fulfilling a duty of his profession. His memory will never die.

Richmann was not by nature foolhardy. On the contrary, he was cautious and well aware of the risks. Fear of lightning, he held, is quite natural and will be overcome only when one understands the phenomenon thoroughly. This in turn requires experiments, and therefore "even the physicist has an opportunity to display his fortitude." Thus his own insight, and his friendship with the great Lomonosov, ensured that Richmann's death was not in vain.

It was almost inevitable that the rush of enthusiasm for taming Jove's thunderbolt would exact a price. Franklin, d'Alibard, Lomonosov, and many others throughout the world had exposed themselves to the same danger as Richmann but had escaped injury. The difference was mostly luck, but Franklin had from the beginning understood something that was dramatically emphasized by Richmann's accident yet took several years to be generally appreciated: Electricity from the clouds is harmless when it has somewhere to go, but dangerous when it finds no exit. Lightning rods attract electricity, but in order to deposit it safely they must be connected by a continuous wire with the ground. In the parlance of the electrician, they must be grounded. For the purpose of experimentation the wire can be interrupted, but after a short gap a path to the ground must be provided. (Franklin's own lightning rod displayed such an interruption: A chime was interposed between roof and ground for signaling the arrival of thunderclouds. When the ringing annoyed Mrs. Franklin while Ben was in Europe, she followed his suggestion to short out the bell with a piece of wire). The thunder machines of Lomonosov and Richmann, on the other hand, were insulated from the ground and therefore lethal. Even the astute Lomonosov, in his description of the tragedy, did not immediately draw the right conclusion. Instead of urging that the machine be grounded, he only cautioned people to stay away from it.

Fear of lightning is natural and wise. Even after its electrical nature was understood, lightning continued to kill, often unnecessarily. For example, it had been common practice for centuries to ring churchbells in order to avert lightning. It was said that "the poor believed that the pious exercise dispersed the evil spirits of the storm, whilst the better sort said it caused some kind of undulation in the air and broke up the continuity of the lightning path." However, a pamphlet against bell-ringing published decades after the invention of the lightning rod still reported that in the previous thirty-three years, 386 church towers had been struck and 103 bell-ringers killed.

Today, in spite of all our science, lightning causes more deaths directly than any other meteorological condition. In the United States it kills about a hundred people and injures twice as many every year. Damage to buildings and ships has been reduced considerably by the introduction of lightning rods, but beyond recommendations for behavior during thunderstorms little can be done to protect people and cattle outdoors. Automobiles and airplanes, because they are made of metal, protect their passengers by conducting the lightning around them. Forests will always be vulnerable to ignition by lightning.

What is this horrid bolt that flings death and destruction amid howling winds, pelting rain, and terrifying thunder, in fiery jagged lines from the heavens down to earth to chastise trembling mortals? What is its cause? Its nature? What are its forms? How does it start? Which way does it travel? How fast? When does it strike?

Zeus, Jupiter, Jehovah, Toth, Thor, and Indra hurled thunderbolts at those they wished to punish. Among American Indians and the tribes of Africa, lightning and thunder were carried by the Thunderbird. All ancient myths agree on the interpretation of lightning as a token of divine displeasure. Today, in stylized form, lightning bolts clutched in the talons of the Jovian eagle on the US dollar bill symbolize war.

Lucretius, the intrepid atheist and materialist, began the long search for scientific explanations of lightning with a series of embarrassing questions:

If it is really Jupiter and the other gods who rock the flashing frame of heaven with this appalling din and hurl their fire wherever they have a mind, why do they not see to it that those who have perpetrated some abominable outrage are struck by lightning and exhale its flames from a breast transfixed, for a dire warning to mortals? Why, instead, is some man with a conscience clear of any sin shrouded unmeriting in a sheet of flame, trapped and tangled without warning in the fiery storm from heaven? Why do the throwers waste their strength on deserts? Are they getting their hand in and exercising their arms? And why do they allow the Father's weapon to be blunted on the ground? Why does Jupiter himself put up with this, instead of saving it for his enemies? Why, again, does he never hurl his bolt upon the earth and let loose his thunder out of a sky that is wholly blue? Does he wait till clouds have gathered so that he can slip down into them and aim his blows at close range? Why does he launch them into the sea? What is his grudge against the waves and the liquid amplitude of the ocean prairies? If he wants us to beware of the flying bolt, why is he loth to let us see it on its path? If on the other hand he intends the fire to strike us unawares, why does he thunder from the same quarter and so put us on our guard? Why does he herald its coming with darkness and mutterings and rumblings? And how can you believe that he hurls it in several directions at once? Or dare you assert that it never happens that several strokes are let fly at the same time? In fact it does happen very often; just as downpours of rain occur simultaneously in many districts, so it must happen that many thunderbolts fall simultaneously. Lastly, why does he demolish the holy shrines of the gods and his own splendid abodes with a devastating bolt? Why does he smash masterly images of the gods and rob his own portraits of reverence with a sacrilegious stroke? Why has he a special fondness for high places so that we see most traces of his fire on mountain tops?

Because in fact, answers Lucretius, lightning and thunder are natural phenomena caused by the crashing together of clouds. He saw lightning as a kind of fire, a reasonable assumption in view of its incendiary power. It wasn't until much later that sparks were found to cause fire without themselves being flames, thus suggesting an alternative explanation of lightning.

Lightning

Experimentation with electricity in the laboratory became possible with the development of the electrical machine in the eighteenth century. Like telescope and camera that determined the course of astronomy, the microscope that ushered in a new age of biology, the analytical balance that separated chemistry from alchemy, and the thermometer that led to thermodynamics, the electrical machine stimulated quantitative investigations of lightning and ultimately of the nature of matter. Unlike today's electronic gadgets, locked magicians' trunks to the uninitiated, the first electrical machines were open to view, easily understood, and pleasurable to the senses. The apparatus of Edward Nairne, master instrument maker in London in the 1770s, represents the best of its kind.

In Nairne's workshop, crowded with the ingenious mechanical paraphernalia of the craftsman and natural philosopher, the electrical machine takes the place of honor. A heavy glass cylinder, about the size of a small cask, is mounted horizontally at chest height on a sturdy frame. At its tapered ends minute cracks, signs of age and use, tessellate the hard black pitch that seals gleaming brass caps to the glass. Stubby axles project from the caps into greased bearings in the frame. The cylinder, in spite of its weight and size, turns easily and smoothly about its long central axis. A belt of round leather, arranged like a bicycle chain, loops around a grooved wheel attached to the cylinder and then around a large cranked drive wheel. The drive wheel, a beautiful solid piece of polished wood, is two feet in diameter, ensuring a high speed of rotation for the cylinder. The entire frame, as large as a table, is made of the best mahogany, fashioned along simple, functional, but gracefully curved lines, lacquered and polished to a warm glow. Its parts, which can be packed away separately in upholstered traveling cases, are fastened together by ornate brass fittings. A long brass handle allows the operator to turn the great wheel that drives the cylinder. The whole device gives an impression of strength and efficiency. Like a loom or a printing press it promises to produce something beautiful and useful in the hands of a skilled practitioner.

Friction between the spinning glass cylinder and a cushion

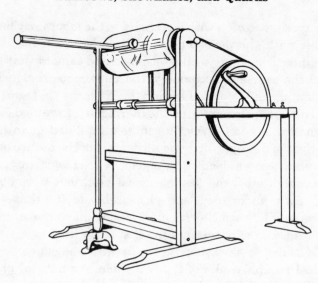

firmly pressed against it by flexible wooden fingers produces electricity. The cushion or rubbing pad, about the size of two adult hands, replaces the hands that had been used for the same purpose a century earlier. At its outermost side, away from the glass, support is provided by a thin mahogany board curved to fit the contour of the cylinder. The soft bulk, the stuffing of the cushion, is horse's hair held in place by a covering of yellow leather, now brittle and gray with use. The glass continually rubbing against the leather produces a small electrostatic charge. Through years of trial and error the electrical artisans of the eighteenth century have enhanced the effectiveness of the machine a hundredfold by use of the proper coating on the leather where it touches the glass. A spoonful of amalgam, a silvery gray coarse powder of zinc, tin, and mercury, sprinkled on the surface of the pad and held on by means of a little lard, transforms the electrical machine from a slightly ludicrous contraption for producing minuscule feeble sparks into a powerful generator of electricity.

What exactly happens when amalgam-coated leather rubs

Lightning

against glass, or a comb through dry hair, or a shoe over a rug in a desiccated room, was not understood two hundred years ago and even today is the subject of speculation and research. As with many apparently simple, common phenomena, the details of the process are subtle, but the outcome—in this case the production of electricity by friction—is easy to observe. Electrification by friction is so reliable that it is used to create huge voltages for powerful nuclear accelerators in modern laboratories.

Once the electricity has been produced on the glass cylinder, it is conducted away from its place of origin to be collected and stored. Benjamin Franklin described this sequence of events by means of a suggestive analogy. He supposed that electricity is an invisible and exceedingly fine fluid that permeates all matter. A normal amount of the fluid renders an object electrically neutral. A deficiency causes a negative charge, while a surfeit represents a positive charge. After the fluid is squeezed out of the rubbing pad in some way, it adheres to the surface of the spinning cylinder, which carries it along. If it were not removed, it would simply return to the pad to restore the proper balance of fluid there.

But the fluid is removed. At the opposite side of the cylinder, away from the rubbing pad, the electrician has placed the prime conductor. Made of burnished brass, mounted at the height of the cylinder on a glass rod in an elegant wooden tripod of Georgian design, the prime conductor consists of a horizontal pipe in the shape of a T. The most curious feature of the whole electrical machine is the line of a dozen fine brass points or needles, soldered to the crossbar of the T and pointing like the teeth of a comb at the revolving cylinder. Although they are separated from it by a fraction of an inch and don't touch it anywhere, the points of the prime conductor receive the electrical fluid in a silent invisible stream. What are they, indeed, but miniature lightning rods that serve to discharge the electrified cylinder?

The stem of the T of the prime conductor serves as the productive end, the spigot, of the electrical machine. Here fine-

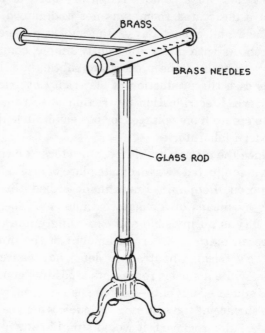

meshed brass chains can be attached to carry the electrical fluid to experiments in other parts of the room. The return connection, or ground, which carries the fluid back to its point of origin after it has performed its task, is provided by another long chain running along the floor, between the curved legs of the prime conductor tripod, under the frame of the machine, and straight up to a little screw in the backboard of the rubbing pad that is being continually discharged at its front surface.

In operation the rhythmic creaking of the frame combines with the sibilant swoosh of the metal bearings, the whisper of the leather belt on wooden wheels, and the crackle of tiny sparks between the links of the chains and on the surface of the cylinder to make a song that is both familiar and welcome to the electrician. Its variations inform him of dry bearings, dirty glass, broken chains, loose screws, slipping belts, excessive moisture, and dust in the air. The machine sings its achievements and its troubles.

Lightning

If the room is dark, the hiss and crackle are accompanied by the twinkling of little sparks, the flashing of bluish crowns at the tips of the discharging points, and occasional streaks of lightning over the surface of the cylinder. These lambent illuminations too are familiar to the operator and reassure him of the good condition of his device. On some days the sounds are wrong, the lights queer, and the temperamental machine won't produce. On others it performs as expected like a faithful old horse, accompanied by its own soft pyrotechnic display of sounds and lights.

To store the electrical fluid, the philosopher uses a Leyden jar, a wonderful Dutch invention. Constructed from an ordinary wide-mouthed glass bottle lined inside and out with tinfoil, the jar is capable of holding a large quantity of electricity in its metal coatings. It is charged by connecting the inner and outer coatings, which do not touch each other, to the prime conductor and the grounding chain, respectively, and it can be carried about or left untouched for many minutes, or even hours, without losing its power.

A dozen Leyden jars hooked together form a battery capable of storing enough electricity to kill a person. Their discharge is the high point in the electrician's art; it is the reward for the hours of work involved in building, assembling, and perfecting the electrical machine. One end of a blunt, crooked brass rod held by an insulating glass handle is gingerly touched to the outside foil of a jar. The other end is brought near a chain that touches the inside foil. Keeping away from all other metal, the electrician peers intently at the shortening gap between the rod and the chain. At a distance of two inches there is a sudden vehement crack, a report sharper and clearer than a gunshot, accompanied by a thick jagged spark, brilliant indigo in color, and so brief that a wink can conceal it. A satisfied smile concludes the experiment, regardless of whether the audience consists of the philosopher alone, or includes colleagues or frightened admiring novices as well. A loud, long spark is the simplest and the most rewarding among hundreds of experiments, perhaps because it represents lightning tamed, Jove's thunderbolt brought under control and wrought out of glass, leather, wood, brass, hair, amalgam, tinfoil, and wax.

A more gentle means of discharging the jars is no less instructive. If a sharp metal point is used instead of the blunted rod, no spark will occur, but the electricity will flow out of the point through the air in an invisible silent stream. It is this discharge through points that underlies the operation of both the prime conductor and the lightning rod. Franklin referred to it at the very beginning of his research in a statement that can serve as paradigm for the power of analogy in scientific research. On November 7, 1749, he wrote in his notebook:

Electrical fluid agrees with lightning in these particulars:

1. Giving light.
2. Color of the light.
3. Crooked direction.
4. Swift motion.
5. Being conducted by metals.
6. Crack or noise in exploding.
7. Subsisting in water or ice.
8. Rending bodies it passes through.
9. Destroying animals.
10. Melting metals.
11. Firing inflammable substances.
12. Sulphureous smell.

The electric fluid is attracted by points. We do not know whether this property is in lightning. But since they agree in all particulars wherein we can already compare them, is it not probable they agree likewise in this? Let the experiment be made.

The experiment was made, and the hypothesis confirmed.

With the years, electrical machines were improved, then replaced by batteries and eventually by electromagnetic generators or dynamos. Today frictional machines are still used by physicists for the production of very energetic electrical discharges. They are cold mechanized monsters made of alumi-

num, steel, rubber, and plastic, but they do not differ in principle from the apparatus of Franklin and Nairne.

As generators became stronger and more reliable, more
was learned about the nature of electricity. Concurrently scientists studied lightning in nature. Fast photography revealed
the startling observation that the usual discharge from a cloud
to the ground is initially invisible. After the first stroke has
touched bottom, a bright visible band forms and travels upward
at great speed along the path previously traversed until the
luminous display reaches the cloud. The double process unfolds
so quickly that it appears to be instantaneous. Nevertheless,
the impression that lightning comes from the clouds to earth
is not without foundation. The zigzag shape of the path resembles an upside-down tree, with branches sprouting from a central trunk. This organic shape suggests growth from the top
down, even though the pattern appears all at once. The mind
uses the familiar simile of a tree more readily than that of
a river system, which moves in the opposite direction, from
the branches toward the trunk.

Whatever its mechanism, lightning is lethal, destructive,
and starkly frightening. Its terror stands in contrast to the
almost trivial simplicity of the rod that averts it. In the whole
compass of technology there are few devices that achieve such
far-reaching beneficial results with such economy, and without
any harmful side effects whatever. Averting the effects of
floods, storms, avalanches, and human aggressiveness is always
expensive and risky. Averting the even more dreadful lightning is child's play. The relief from fear of lightning explains
Franklin's almost divine status in France, which enabled him
to score unprecedented diplomatic successes.

Franklin's invention, like many good ideas that are especially simple, is timeless. The lightning rods on modern buildings are almost indistinguishable from those of their inventor.
A slight variation is often seen from the highway. High-tension
power lines, successors of the brass chains of the eighteenth-
century electricians, themselves need protection against the
superior power of lightning; they receive it from thin wires

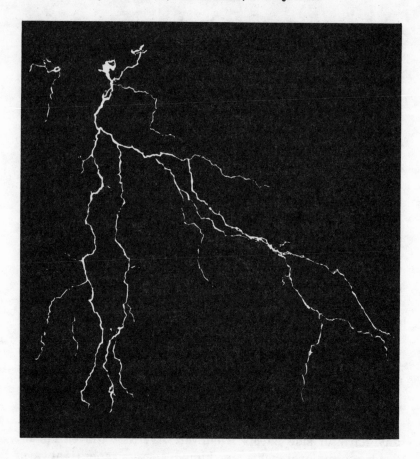

strung from tower to tower above all the other cables. Thus
an outgrowth of Franklin's invention guards the fruits of his
discoveries.

Franklin's theory has not fared so well. His conception
of electricity as a fluid, which elevated the study of electricity
from a curiosity to a science, was natural for the period. Other
phenomena, such as heat, were similarly explained by reference to invisible fluids. Such electrical words as *current, flow,*
and *capacity* attest to the power of the analogy and will remain
in the language, but today the model of electricity is no longer

that of a fluid. The modern notion was actually anticipated by Franklin, but not developed. At the very beginning, before his fluid theory had been broadcast and debated, he speculated on the deeper nature of the fluid itself and wrote: "The electrical matter consists of particles extremely subtile, since it can permeate common matter, even the densest metals, with such ease and freedom as not to receive any perceptible resistance." The particulate nature, the graininess, the atomicity of electricity was proposed by Franklin long before a test of the idea became possible. What prospered instead was his fluid theory.

Natural electricity is generated in the clouds. "Crashing together of clouds," Lucretius called the process, but modern investigators focus more on the collision between individual drops. The science of electricity began with an experiment on charged clouds, performed by an American. A century and a half later, another experiment, also on clouds and also American, put an end to the old theory and established the modern view of electricity.

The modern experiment, by Robert A. Millikan, began as a measurement of the average charge on a small artificial cloud from an atomizer. It quickly became clear that more accurate results could be obtained by considering individual drops, and that oil, because of its slow evaporation, works better than water. Millikan measured the electric charges on thousands of oil drops, one at a time. The result of this tedious program was a resounding triumph of atomism over its ancient rival, the belief in the continuum: The charges did not occur in arbitrary amounts but always in multiples of one smallest unit, the elementary charge. Electricity is not a fluid but a collection of a vast number of individual, identical, "extremely subtile" particles called electrons, each carrying an elementary charge. The currency of electricity is not measured in liquid gold but in a shower of coins.

Georg Richmann died while trying to measure the electrical charge in a cloud. Robert Millikan rendered the endeavor safer and more reproducible by bringing it into the laboratory and confining himself to a single drop. In the process he found a way to measure nature's elementary charge. But the quest is not over. Today there are theoretical reasons for believing that there might exist charges that are even smaller than the smallest found by Millikan. Whether they can be found is a question for experiment. Two papers in the *Physical Review Letters,* which reports the latest and most sophisticated research, are entitled "Results of a Search for Fractional Charges on Mercury Drops" and "Observation of Fractional Charge of $(1/3)e$ on Matter." The jargon means that Millikan's experiment has been refined and repeated, in one case with no new results, and in the other with evidence for a charge one third that of the electron. The contradiction between the two reports promises future excitement, for science feeds on contradiction, puzzlement, uncertainty, and doubt.

The Compass

Like a long medieval Chinese landscape painting, the history of physics unfolds before the mind's eye. Steep snowy mountains, dark mysterious valleys shrouded in mist, wild forests, and peaceful lakes are connected by meandering rivers and torturous rocky paths that carry the viewer through time and space from scene to scene as the scroll is unrolled from right to left.

In the distance among the craggy peaks, the great laws are set down in exquisite calligraphy. The language, instead of Chinese, is mathematics. The symbols, instead of ideograms, are Arabic numerals and letters from a variety of alphabets, joined by computational signs such as the equality and the square root. Like Chinese characters, the equations are common to many spoken languages and thus transcend them. Newton's laws of motion and of gravity stand out majestically. A little further on, Maxwell's four graceful equations describe

the phenomena of electricity and magnetism in concise nota-
tion. They are followed by the laws of thermodynamics, the
rules of special relativity, Einstein's general theory of relativity, the unsolved equations of fluid flow, Schrödinger's equation
of quantum mechanics, and finally the modern descriptions
of particle physics that incorporate many of the previous laws
in densely packed formulations. Between the mountains and
among the equations, great clouds of fog and swirling mists
represent our ignorance of vast portions of the physical world.

The foreground of the painting is crowded with famous
scientists and their attributes. Here's Aristotle, clad in a toga,
gazing at the clouds as he composes his treatise on meteorology;
there, in a monastic cell, Theodoric finds a rainbow in a glass
globe filled with water; barely visible in the New World jungle
that hides curious Indians; Thomas Harriot, servant to Sir
Walter Raleigh, kindles a fire by means of burning glass; Johannes Kepler, trudging across the bridge in Prague, catches
a snowflake on the sleeve of his fur-lined coat; Galileo Galilei
sits attentively in the Cathedral of Pisa, timing the swing of
the great chandelier with his pulse; Sir Isaac Newton, bewigged
and proper, measures the spectrum projected on the wall of
his study; Benjamin Franklin cranks an electrical machine;
Madame Curie stirs a pot of tar; Ernest Lawrence adjusts his
little cyclotron. They all seem intent on watching, even as
they invent explanations of what they see, combining observation with reasoning in the characteristic manner of Western
science. Only Albert Einstein, puffing peacefully on a pipe in
his little sailboat, appears to be lost in dreams. His images
are in his thoughts, his experiments employ the apparatus
of the mind.

Toward the right, earlier in time, the crowd of notables
thins and grows dimmer. One of the figures, in the central
medieval region of the scroll, seems to be more at home in
the landscape than all the others. Sitting on the floor of a
terrace overlooking an ornamental garden, wrapped in a flowing robe with wide belt, his long hair piled in a bun on top
of his head, Shen Kua, a Chinese astronomer and polymath,

contemplates the object in his hand. Close inspection reveals the device to be a compass. It is a piece of technology of astonishing simplicity, power, and mystery. To Shen Kua falls the honor of holding it, not as the inventor, but as the author, in A.D. 1088, of the first recorded description of the compass.

Over the centuries the construction of this wonderful instrument has been improved and standardized to its present state of perfection. A glass cover plate is sealed to the housing made of some nonmagnetic material such as brass. The magnetized steel needle, its northern half blue, the southern tip silver, balances on a vertical pin projecting upward into a little indentation in the center of the needle. In use, the glass does not touch the needle, but is close enough to prevent its slipping off the pivot even when the compass is turned upside down. The card inscribed below the needle indicates the eight principal directions, or points, by means of a stylized rose, as well as 360° marked off starting with zero in the north and proceeding eastward. The construction is as simple as it is rugged and reliable.

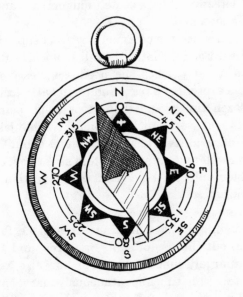

As an aid to navigation the compass made possible the voyages of discovery. Without it, direction can be found in the middle of the ocean only on clear nights when the North Star is visible. During the day the sun is of no help without both an ephemeris that charts its course through the sky and an accurate clock. In cloudy weather, among unfamiliar patterns of waves and winds, all directions look the same. The unerring dependability of the compass must have provided great comfort to the nautical adventurers of the fifteenth and sixteenth centuries. Later it led explorers across the poles and into the interiors of the continents. Today it guides as surely as it did when it was invented.

Not only in history but also in science the compass occupies a special place. It is the prototype of all the meters, dials, gauges, and pointers that surround us at home, in cars and airplanes, at the factory, and in the laboratory. Two essential elements of analog meters first combined in the compass: the fixed graduated scale and, above it, the movable needle that responds automatically to an external stimulus. The compass led directly to the galvanometer, which measures electrical current by detecting its associated magnetism, and thence to all the other meters.

The compass itself is common now, a cheap trinket found in dime stores. For all that, it is still magical. After the case is placed on a flat surface the needle jiggles a little bit to find its balance, oscillates back and forth with decreasing amplitude a dozen times, and then stops dead, pointing north and south as if it were stuck in some invisible jelly. Anyone with a sense of curiosity is struck by wonder about the phenomenon and can well understand Einstein's reaction, as described in his autobiography in the context of a reflection on the process of thinking:

> I am convinced that thinking takes place for the most part without the use of signs or words, and that it is in addition largely subconscious. For how else would it happen that we sometimes spontaneously marvel at an experi-

ence? This marveling seems to occur when an experience comes into conflict with the totality of concepts that is already sufficiently fixed within us. When such a conflict is felt strongly and intensively it reacts back upon our thoughts in a decisive way. The development of the world of thoughts is in some sense a continuous flight from marvel.

A marvel of such a kind I experienced as a child of four or five years when my father showed me a compass. That the needle behaved in such a determined way did not at all fit into the nature of those events that could be accommodated in my subconscious world of concepts (where actions were related to "touch"). I still remember—or at least believe I remember—that this experience made a deep and lasting impression upon me. Something deeply hidden had to be behind these things. What people see from infancy, to that they do not react in such a way. They don't marvel at the fall of a body, at the wind and the rain, nor at the moon and the fact that it doesn't fall down, nor at the difference between living and non-living things.

Einstein's point is that the mystery of the compass stems from its apparent violation of intuition formed by experience. Almost all the forces found in everyday life work through "contact" or "touch." The hand grasps the book and the pot in order to move them, the foot kicks the ball in order to propel it, the motor engages the wheel in order to turn it. To be sure, there are causes that seem to have effects without contact—the force of gravity, the force of the wind, the transmission of sound. But these, says Einstein, are so familiar that they enjoy uncritical acceptance. The compass, by contrast, is unusual enough to be remarkable.

What causes it to turn? Is there *something*, after all, where empty space should be, tugging on the needle to force it into alignment? Is there some invisible jelly?

The three scientific effects that underlie the operation of

the compass are magnetic attraction, magnetic polarity, and the magnetic field of the earth. To these are added technical developments such as the use of steel for the pointer, the suspension of the needle, and the graduated scale. The first of these is very much older than the compass. The ancient Greeks knew of a certain rock, called lodestone or magnet, that attracts iron.

Beyond noting its existence, Western scholars paid little attention to the magnet. For centuries while mathematics matured and astronomy became the queen of the sciences, magnetism languished as a footnote. To make things worse, it was often confused with electricity, the attractive power of amber. In China, on the other hand, the lodestone was used and its properties harnessed in a practical invention—the compass.

The tale of the development of the compass has been unraveled for Western eyes in the panoramic work of historian of science Joseph Needham. Comparable in scope to Sir James Frazer's *Golden Bough* and Arnold Toynbee's *Study of History*, Needham's *Science and Civilization in China*, begun in 1954 and still unfinished, is a vast and ambitious undertaking. Its dozen volumes will recount the whole story of Chinese science and technology, in their relation to the arts and to society, from the earliest records to the seventeenth century.

While it is difficult to compare Chinese with Western science because the presuppositions and questions were often so very different in nature, it is possible and instructive to compare technologies. A ship, after all, is a ship, and a stove a stove, even if they differ in design and manner of operation. Needham traces the evolution of countless Chinese devices and techniques as far back as his voluminous sources allow. He spends a great deal of effort in determining the dates of their first introduction in China and in Europe in order to establish priority.

The point of this exercise is not an invidious comparison of the two cultures, but a process Needham likens to titration. Titration is the chemical analysis of a solution by a measurement of how much of another substance must be added in order to change the color of the solution. The solutions to be analyzed in this case are Western and Oriental civilizations, and the color changes are great social upheavals. Each society has been deeply affected by the introduction of discoveries and inventions from the other. By a careful study of these mutual influences, Needham seeks to add to the understanding of the history of both. Like most sinologists, he is fascinated by the question "Why, if China led Europe in many aspects of technology during the first fifteen centuries A.D., and experienced no dark ages, did modern science develop only in the West?" But instead of losing himself in speculation about this elusive problem, Needham uses the question as a guiding principle for plunging into the exhilarating task of finding out just what Chinese science did accomplish.

The compass plays an important role in this process of titration. Since the Renaissance, the West has acknowledged three Chinese inventions of incalculable social significance: the compass, the art of printing, and gunpowder. In addition to these, Needham found a hundred more, including mechanical clocks, cast iron, sequential arch bridges, the stern-post rudder, fore-and-aft sailing, and quantitative cartography. But from all of these, Needham singles out the compass as China's greatest contribution to physics.

The story of the compass began centuries before the Christian era with the art of divination. Among the tools of the magician were small symbolic objects, ancestors of our dice, thrown or arranged on the ground. They had astrological significance and one of them represented the constellation that is known to every child, even in the West—Ursa Major or the Big Dipper. Its function, like that of the compass, is to point to the north. Eventually the piece representing the Big Dipper evolved into the shape of an actual dipper or spoon consisting of a round bowl with a short stubby handle, like a modern Japanese soup-spoon.

A fateful step was taken by an ancient diviner who decided to fashion his pieces of the magical material magnetite, the stuff of the lodestone, instead of the more usual wood or clay.

This trick enabled the symbolic objects to move about under their own power, rearranging themselves mysteriously without human intervention. The inventor could even manipulate them without contact by means of a strong magnet held under the table. This idea has survived into today's novelty shops, where plastic dolls pirouette on smooth mirrors, driven by concealed magnets.

As a delightfully surprising aside, Needham relates the divining symbols to the game of chess. That chess should be astrological in character is more natural in China than in the West because Chinese astrology was from the beginning judicial rather than horoscopic. Its function was to foretell and interpret the fates of rulers and states, rather than those of individuals. The battle element of chess, according to Needham, developed from a technique of divination in which the ever-contending forces of yin and yang in the universe were pitted against each other. Later the magical procedures were transformed into a recreational game in India.

A diviner's board, called a *shih*, complemented the movable pieces. It consisted of a lower, square earth plate surmounted by a disk, the heaven plate, which rotated like a lazy susan. Both were marked by astronomical signs and by complicated symbols of divination. Often the heaven plate was engraved with the pattern of the Big Dipper. Needham's book contains a photograph of a fragment of the heaven plate of a first-century Korean *shih*. The seven stars of the Great Bear are instantly recognizable even though they are shown backward, as though seen from a vantage point above and beyond the celestial sphere. (A miniature human being, standing on the earth plate and below the heaven plate, would see the constellation properly above him.) In the square earth plate Needham finds the ancestor of the chess board, in the round heaven plate, the prototype of the compass dial.

The crucial event in the development of the compass was the replacement, some time during the first six centuries A.D., of the heaven plate, with its engraved Dipper, by a magnetic spoon. If all the surfaces were highly polished, the spoon, balanced on the lowest point of its bowl, would act as a compass and align itself with the earth's field. Thus, serendipitously, the "south-pointing spoon," or at times the "north-pointing spoon," was discovered. Although no examples of this device have been found, models have been constructed according to the numerous written descriptions, and found to exhibit the expected behavior.

The south-pointing spoon was followed by the south-pointing wooden fish, which incorporated a magnet and overcame the troublesome problem of frictional drag by floating on water. Later it was found that steel could be magnetized, either by contact with a lodestone or by heating, and in the seventh or eighth century a steel needle, suspended by a silk thread, replaced the spoon and the fish.

In 1088, almost exactly a century before the first European mention of the compass, Shen Kua wrote in his *Brush Talks from Dream Brook:*

Magicians rub the point of a needle with the lodestone; then it is able to point to the south. But it always inclines slightly to the east, and does not point directly at the south. [It may be made to] float on the surface of water, but it is then rather unsteady. It may be balanced on the finger-nail, or on the rim of a cup, where it can be made to turn more easily, but these supports being hard and smooth, it is liable to fall off. It is best to suspend it by a single cocoon fibre of new silk attached to the centre of the needle by a piece of wax the size of a mustard-seed—then, hanging in a windless place, it will always point to the south.

Among such needles there are some which, after being rubbed, point to the north. I have needles of both kinds by me. The south-pointing property of the lodestone is like the habit of cypress-trees of always pointing to the west. No one can explain the principles of these things.

Further on we find:

When the point of a needle is rubbed with the lodestone, then the sharp end always points south, but some needles point to the north. I suppose that the nature of the stones are not all alike. Just so, at the summer solstice the deer shed their horns, and at the winter solstice the elks do so. Since the south and the north are two opposites, there must be a fundamental difference between them. This has not yet been investigated deeply enough.

These are the first unambiguous references to the compass. Shen Kua, like most intellectuals of his time, was a civil servant in the Imperial bureaucracy. A brilliant career, concerned with mathematics applied to cartography, calendrics, astronomy, and even finance, ended suddenly in impeachment precipitated by a jealous rival. Shen retired to a garden estate that he had dreamed about before he ever saw it, and therefore named Dream Brook. There he pursued his muse, writing on mathematics, administration, astronomy, warfare, painting, tea, medicine, poetry, music, and countless other topics. His greatest work on science and technology is called *Brush Talks from Dream Brook* because, he said, he had "only my writing brush and ink slab to converse with." This work, a congeries of six hundred recollections and observations, has earned him a great reputation as polymath, and comparison with such latter-day luminaries as Leibnitz and Lomonosov.

The aphoristic nature of Shen's writings is characteristic of the science of his day. Missing was the conviction that all natural phenomena could be understood by the systematic application of reason. Nature was considered too rich and subtle to yield to the kind of analysis that is the hallmark of modern science. "No one can explain the principles of these things," Shen says humbly of the compass. But metaphors and analogies were as important to him as they are to the popularizer of science and, in a more refined way, to the modern scientist. Where Shen in the eleventh century compared the compass to cypress trees bent in the wind, Niels Bohr, in the twentieth, saw a solar system in the structure of the hydrogen atom. The difference is that Bohr recognized a common cause for the two disparate phenomena he compared, while Shen did not.

The second sentence in Shen's description of the compass struck Needham like a bolt out of the blue. The observation that the needle does not point exactly north and south refers to the phenomenon of magnetic declination. The deviation differs from place to place on earth, and is a matter of life and death for mariners. In the West, tradition has it that Christopher Columbus discovered declination with considerable consternation on his voyage of 1492. There is some evidence for slightly earlier European observations, but six hundred years before Columbus, the Chinese diviners were well aware of it.

Both before and after Shen Kua, the chief users of compasses, besides sailors, were geomancers. Their task was to bring the houses of the living and the tombs of the dead into harmony with the winds, the waters, and the spiritual breath of the earth. Thus magnetic science grew out of the pseudoscience of geomancy in China, just as in the West astronomy derived from astrology and chemistry from alchemy.

In Europe, magnetism finally achieved an important status in science through the work of Queen Elizabeth's physician William Gilbert. His book *On the Magnet,* published in 1600, was justly acclaimed as definitive by the foremost scientists of the day, including Galileo and Kepler. It is erudite and complete in its quotations of ancient authors, but heaps scorn on those who "having made no magnetical experiments . . .

constructed certain ratiocinations on a basis of mere opinions, and old-womanishly dreamt the things that were not." Thoroughly empirical in spirit, the book may be regarded as the harbinger of modern science.

Gilbert correctly placed the invention of the compass in China. (His belief that Marco Polo introduced the compass to the West is contradicted by European descriptions a century before Marco Polo's voyage.) Both as an instrument for making manifest the invisible influence of the magnet, and as a quantitative measuring device, the compass played a central role in Gilbert's experiments.

One of Gilbert's most important achievements was the explanation of why the compass points north and south. Earlier writers had tried everything from analogies with cypresses to the influence of the pole star. Gilbert, on the other hand, fashioned a terrella, a little model globe made of lodestone, and showed that its effect on the compass accurately mimics the effect of the earth. He concluded that the earth is, in fact, itself a giant magnet.

Two puzzles emerge from this theory. The first is merely semantic: Why is the north pole of a compass needle attracted to the south pole of a magnet, but to the *north* pole of the earth? The answer is simply that either the earth's poles, or all the magnets in the world, are misnamed. Since it is easier to change two names than a million, scientists tacitly acknowledge that the earth's north pole is really a magnetic south pole and vice versa.

The second puzzle concerns declination. If the earth is a gigantic magnet, why does the needle not point truly north and south? The answer, provided by Gilbert, is that there is no reason for the geographic and magnetic poles to coincide. The geographic poles are the ends of the axis about which the earth turns. The magnetic poles are the ends of a hypothetical magnet inside the earth, and this magnet is not lined up with the axis of the earth. In fact, the magnetic poles wander. One is currently located on a Canadian island, fourteen degrees from the North Pole, the other off the coast of Antarctica, twenty-four degrees from the South Pole.

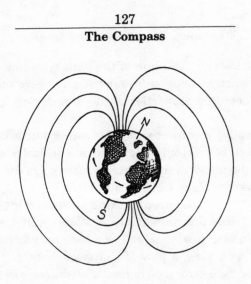

It is in the nature of science that Gilbert's explanation of the operation of the compass raises new questions. What makes the earth behave like a huge magnet? More fundamentally, how does any magnet exert its influence on another magnet? The first question has led to the whole science of geomagnetism, to speculation about the magnetic properties of the molten metallic core of the earth, to theories of the terrestrial dynamo, and to current research and controversy. The second question leads to electromagnetic theory and to atomic physics.

The relationship between magnetism and electricity, for centuries confused by careless observation, was clarified by Gilbert. He firmly and convincingly divorced the two concepts. The affinity of chaff for rubbed amber has nothing to do with the attraction of iron to the lodestone. An electrical charge, even a very large one produced by an electrical machine, has no effect on a compass. And yet—a remnant of suspicion remained that there might be a connection. In 1735 an article appeared in London under the title "Of an Extraordinary Effect of Lightning in Communicating Magnetism." It described how a bolt of lightning had rendered a box full of knives and forks strongly magnetic. The electrical nature of lightning was not firmly established until fifteen years later, so the hint came too early. After the invention of the battery, which could supply steady electrical currents, experiments were made to discover the magnetic properties of the new device. There were none.

But finally, unexpectedly, the connection was made. It turned out to be not only deep and scientifically significant, but also of the utmost importance to society. On July 21, 1820, Danish physicist Hans Christian Oersted reported the strong deflection of a compass needle near a wire carrying a current. Here at last was the connection. A static charge, whether it resides in an electrical machine or a battery, has no magnetic virtue. But a current, a flow of charges, whether it is carried by a wire or by a stroke of lightning, behaves like a magnet. Oersted's discovery immediately captured the imagination of scientists throughout the world and led quickly to the realization that the relationship is reciprocal. Moving charges create a magnetic field, and moving magnets create electrical currents. These two observations are the basis not only of generators and motors, but also of telephones, radio, TV, and all the trinkets of our information-sodden civilization. Shen Kua's

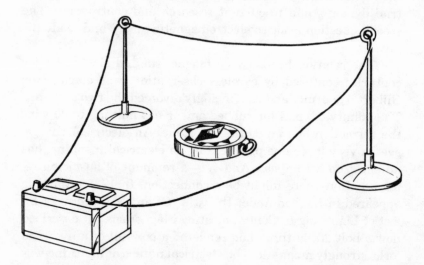

toy, though innocent in appearance, bears a heavy burden of responsibility.

Like Einstein, who marveled at the power of the compass to find its proper direction, and Needham, who was astonished to discover a description of magnetic declination in the eleventh century, I too had a memorable experience with a compass. When I was about fourteen my father told me about Oersted's discovery. He explained that if a current is passed above a compass, the needle will always line up perpendicular to the wire. I understood this, but then a disagreement arose. Imagine a compass lying on a table, and just above it a straight wire carrying a current northward. From symmetry, and from what I had just learned, I confidently predicted that the needle will turn to point either east or west, depending on its exact starting position. My father, however, insisted that it will always point west, never east. I did not believe him. Since the current is going north, straight up the middle, so to speak, there is nothing in the experiment to distinguish east from west. Both directions must be equally acceptable to the needle. I was so sure of my intuition, based on symmetry, that I flatly contradicted my father—who has a Ph.D. in physics. He was understandably displeased, but this conversation contributed to my own decision to study physics.

Of course my father was right, but my intuition was right too. If the situation were as symmetrical as it looks, the needle really couldn't exhibit a preference for the west. There is, in fact, an asymmetry, but it is deeply hidden in the nature of magnetism, and invisible to the eye. The needle of the compass derives its power partly from tiny elementary current loops in the atoms of the material. Some carry current clockwise, others counterclockwise. It happens that in a magnetic material there are more of one kind than of another, and this preference for motion in one direction confers on the needle an invisible handedness that spoils its apparent symmetry.

The fascination of the compass needle continues. It has been known since long before Gilbert that if a compass needle is broken in the middle, the result will not be two isolated poles, but rather two short needles with two poles each. Magnetic objects always combine a south pole with a north pole. There has never been such a thing as a single north pole, or south pole. Nature abhors monopoles.

And yet, in May 1982, the *Physical Review Letters* carried a paper entitled "First Results from a Superconductive Detector for Moving Magnetic Monopoles" by Blas Cabrera, which described the observation of a monopole. If it is verified, it will upset much that has been learned about magnetism in the last two millennia. It may also make possible the first significant modification in the design of the compass since Shen Kua: a needle with a north pole, but no south pole.

Snowflakes

On a snowy day in the winter of 1609, Johannes Kepler, astronomer and Imperial Mathematician to the court of Rudolph II in Prague, pushed aside his books of computations and began to compose a letter:

> I am well aware how fond you are of Nothing, not so much for its low price as for the sport, as delightful as it is witty, that it affords your pert sparrow; and so I can readily guess that the closer a gift comes to Nothing the more welcome and acceptable it will be to you . . .
>
> In such anxious reflections as this, I crossed the bridge, embarrassed by my discourtesy in having appeared before you without a New Year's present, except in so far as I harp ceaselessly on the same chord and repeatedly bring forth Nothing: vexed too at not finding what is next to Nothing, yet lends itself to sharpness of wit. Just then

by a happy chance water-vapour was condensed by the cold into snow, and specks of down fell here and there on my coat, all with six corners and feathered radii. 'Pon my word, here was something smaller than any drop, yet with a pattern; here was the ideal New Year's gift for the devotee of Nothing, the very thing for a mathematician to give, who has Nothing and receives Nothing, since it comes down from heaven and looks like a star.

Back to our patron while the New Year's gift lasts, for fear that the warm glow of my body should melt it into nothing.

Kepler was at the height of his career. At age thirty-eight he had finally finished a six-year grind of calculation leading to the shattering realization, painfully forced on him by careful observation, that planets travel neither in perfect Aristotelian circles, nor in Ptolemaic circles imposed upon circles, but in ellipses. He was an intense and handsome man with a long straight German nose, bordered by deep furrows that bear witness to a troubled youth and tumultuous years of early adulthood. The bridge referred to in the letter is the ancient Karlsbrücke over the Moldau, linking palace and town by sixteen gothic arches. The patron, from whom Kepler has just come and to whom he hurries back with his ephemeral gift, is Johan Matthäus Wacker von Wackenfels, Counsellor at the court, lawyer, diplomat, intellectual, poetaster, and lover of literary trifles. The reference to Nothing is a pun: *Nix* means "snow" in Latin but it also means "nothing" in colloquial German. The allusion to the sparrow is obscure. The whole passage opens Kepler's wonderful little essay on the problem of the shape of snow crystals entitled "A New Year's Gift or On the Six-Cornered Snowflake."

Kepler was of course not the first to recognize the symmetry of ice crystals. Unknown to him, Chinese scholars had noticed it as early as 135 B.C. In the West, universal naturalist Albertus Magnus wrote about it in 1260, and others followed, but none with the enthusiasm of Kepler.

Snowflakes

The problem is posed with exceptional clarity:

> Our question is, why snowflakes in their first falling, before they are entangled in larger plumes, always fall with six corners and with six rods, tufted like feathers. . . . There must be some definite cause why, whenever snow begins to fall, its initial formation invariably displays the shape of a six-cornered starlet. For if it happens by chance, why do they not fall just as well with five corners or with seven? Why always with six, so long as they are not tumbled and tangled in masses by irregular drifting, but still remain widespread and scattered?

The identification of a well-posed problem is always a welcome event in scientific research. It is surprising to find it at the very awakening of modern science, so near the end of the Middle Ages that preferred disputation over hypotheses to simple questions about observations.

The observation underlying the question is correct. Microscopy and photography have enlarged and preserved the images that caught Kepler's eye. Of the thousands of drawings, photographs, and plastic casts that have been made since then, almost all exhibit hexagonal symmetry. Experts have proposed complicated schemes of classification for the multitude of pretty shapes, complete with the obligatory alphanumeric labels and ugly names, but the symmetry remains.

The classic snowflake is a little star of great beauty. Usually a regular hexagonal flat plate, often with minute concentric six-sided rims and ridges parallel to the edges, forms the center. From each corner (never from a side), an arm or radius—a rod, as Kepler put it—juts out to a distance that may be many times larger than the diameter of the central plate. At irregular intervals branches grow from the radii. The key to the structure of snowflakes is the observation made by Descartes that the branches grow only in directions parallel to the adjacent arms, always at 60° to their stems. Each little branch has parallel sides and a blunt point. Sometimes the

branches grow progressively shorter along the arm, so that the whole flake looks like six Christmas trees joined at their bases. In other cases the branches are of equal length and the arm terminates in another hexagonal plate instead of a point, giving the flake the appearance of a decoration on the chest of a Czarist general. Sometimes the branches are so close together that they fuse and the flake becomes a six-petaled flower. Occasionally the arms are thick and bare with no branches at all, giving them a mechanical look, like the product of a giant press that has flattened a game of jacks. Other flakes are adorned with such exquisitely fine whiskers that they resemble down. The most ornate snow crystals sport twigs on the branches of the radii, always growing at angles of 60° to each other.

Besides these feathery stellar shapes there are needles, columns, columns with end-plates, and many others. The remark that no two are ever alike is not so much a statement about nature's abundance as it is about the precision with which the comparison is made. At the atomic level no two grains of sand, no blades of grass, no drops of rain, no buttons, no pins are identical even though they look alike from a distance. The same is true for ice crystals. Whether or not it is possible to find two snowflakes that appear identical to the human eye depends on the precision of observation, and is not a useful question.

After pointing out the essential six-sidedness of snowflakes, Kepler proceeded with the quest for the reason, and immediately set out in the wrong direction. With the blast of a trumpet that signals the charge he began: "The cause was not to be looked for in the material, but in an agent." He based his faith on the incorrect notion that vapor droplets, being spherical, cannot contain the seeds of shapes and patterns. This assumption was characteristic not only of Kepler, but of a whole tradition in science: the rejection of atomism. Instead of seeking the causes of material phenomena deep down in the structure of matter in ever-smaller units and subunits, Kepler looked for external agents, forces, or principles, powers that were just as impalpable as atoms were in his time, and as quarks are in ours.

In search of order in the universe, the natural philosopher first looks for enlightening analogs, for related examples, for similes and metaphors. Kepler wasted no time in opening a long parenthesis on other symmetric structures in nature. One of the most beautiful is the bee's honeycomb, which shares, with snowflakes and bathroom tiles, a hexagonal symmetry. In fact, however, the walls of the cells are rounded, not angular, on the inside where the bee lives, so that the architectural plan could also be thought of as a pattern of snugly touching circles. Pennies on a table, arranged and pushed as close to each other as possible, assume the same pattern. The little triangular spaces between the pennies represent the areas filled by wax in the hive, while the pennies themselves represent the living quarters. The hexagonal appearance derives from the fact that each penny and each apiary tunnel is touched by six neighbors. This arrangement produces the most useful space for the smallest wall area, and hence consumes the smallest amount of wax. If the cells were triangular or

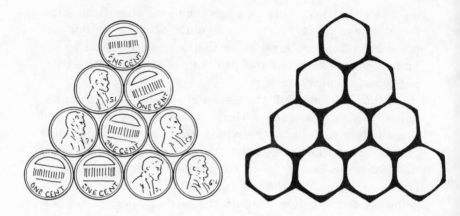

square in cross-section, the bees would have to produce more wax per dwelling unit, and in addition would have to put up with sharp internal corners. The hexagonal arrangement is the most efficient.

From honeycombs, Kepler went on to three-dimensional structures and in particular to pomegranates. If each seed were originally spherical, and if then the whole fruit were to swell from the inside without expansion of the rind so that the seeds were squeezed together, what shapes would they assume? The answer, according to Kepler, is a rhombic dodecahedron, a solid with twelve diamond-shaped faces. Each seed is surrounded by six others in a plane, as well as three above and three below. Just as the hexagon is the most efficient plane-filling figure, so the rhombic dodecahedron is the most efficient volume-filling polyhedron. Most remarkably, the ends of honeycombs, where bees working from opposite sides of a vertical wall of a hive meet, are also constructed in the shape of rhombic dodecahedra. The three little rhombs that close each tunnel are a pretty sight to discover in jars of unfiltered honey.

The mathematical proof that the angles and planes preferred by bees are in fact the most efficient way to save wax, even at the end walls, had to await the invention of calculus long after Kepler. But how do bees do it? Since they don't calculate, their skill must be ascribed to an outside agency. In the eighteenth century the French Academy officially denied bees the geometric intelligence of Newton and Leibnitz, but concluded that they must obey divine orders that are in keeping with the laws of mathematics. A more naturalistic interpretation was given by Charles Darwin, who spoke of the architectural ability of bees as "the most wonderful of known instincts" and added that natural selection in this instance has come to an end, for it could not lead "beyond this stage of perfection."

Kepler closed his parenthesis on six-sided symmetry in the plane and in space by a consideration of ways of stacking cannonballs. He distinguished two different methods, today called hexagonal and cubic close packing. The former leads to packed rhombic dodecahedra, the latter to a simple stack of cubes. They represent two efficient ways of packing identical spheres into a given volume.

A paradoxical illustration of close packing can be observed on a walk along the seashore. As the foot presses on the sand where the falling tide leaves it firm, the area immediately surrounding the foot becomes momentarily dry. Most people, when asked whether the sand is compressed by the foot, would unhesitatingly answer "Yes!" But that answer, as first pointed out by Osborne Reynolds at the meeting of the British Association at Aberdeen in 1885, conflicts with the observation. If a wet rug is compressed by the weight of a person walking on it, the footsteps are surrounded by puddles, not dry patches. The conclusion is that sand is *not* compressed by the pressure of the foot. Rather, its natural close packing is disturbed by the external force, causing a disordering and hence an expansion by the opening of interstices through which the water escapes. For anyone who remembers this argument, it is impossible to walk along the seashore without watching the phenomenon. Thus natural philosophy, that excellent alloy of observation and rational analysis, leads to a deeper appreciation of the world.

Kepler's distinction between different types of close packing of spheres represented a pioneering contribution to the science of crystallography, but because atoms were not part of his cosmology it led him nowhere. When he tried to use cannonballs as models for vapor droplets that condense to make a snowflake, honesty compelled him to list insurmountable objections to each conjecture. About halfway through the essay, with the end of the year approaching rapidly, he admitted to a time-honored schoolboy ruse for stretching the length of the essay: "I shall push this notion as far as it will take me, and only afterwards shall I test its truth, for fear that the ill-timed detection of a groundless assumption may perhaps prevent me from fulfilling my engagement to discourse about a thing of Naught." Such candor is as typical for Kepler as it is rare in general.

All material causes were rejected until only one possibility remained: "the Creator's design." The agent for carrying it out was the *Facultas Formatrix,* the Formative Faculty. This vague morphogenic principle was responsible for all shapes in nature, whether inorganic as those of snowflakes, planetary orbits, and the paths of light rays, or organic as those of plants and animals. Today, appeal to the formative faculty appears to be an admission of defeat. After posing the question so carefully, Kepler was reduced to answering that snowflakes are hexagonal because it is in their nature to be so. There is no unifying or predictive power in the conclusion. Kepler was poised precisely on the threshold between modern science and ancient scholasticism, and his New Year's gift was balanced, in the clarity of its question and the obscurity of its answer, between the two.

The scientist in Kepler sensed the inadequacy of the argument, for in the end he made a last feeble attempt to pass the buck. Knowing that different salts crystallize into different shapes, Kepler wondered whether the seeds of the shapes of snow crystals were not perhaps carried by salts dissolved in the water, and whether these salts did not, in fact, exhibit hexagonal symmetry. So he challenged the chemists to investigate the conjecture.

Snowflakes

The snowflake has continued to charm and baffle. Long after Kepler the chemical formula for water, the familiar H_2O, was determined. Then, about fifty years ago, the arrangement of the atoms in the molecule, with two little hydrogens stuck onto a big oxygen like ears on Mickey Mouse's head, was discovered. The secret of the snowflake lies in the angle subtended by the hydrogen atoms at the center of the oxygen atom: It is almost 120°. As a consequence of this structure, ice forms a stiff hexagonal lattice. A crystal, enlarged by the condensation of water on its surface, preserves the original symmetry. Thus the key to the shape of snowflakes finally is found in the atomic structure of water.

How prosaic that sounds, how simple, and how inadequate. The symmetry of the parts doesn't necessarily imply symmetry of the whole. Given a million hexagonal bathroom tiles, it is possible to fit them together in the form of a reasonably faithful silhouette of Abraham Lincoln without a trace of large-scale hexagonal symmetry. Similarly, the atomic structure of water makes the symmetry of snowflakes possible, but not necessary. Atomism alone is not sufficient to answer Kepler's question; other considerations must be adduced.

Modern descriptions of the growth of a snowflake are long, complicated, and tentative. The speed with which water solidifies on contact with ice depends on the temperature, on the amount of water vapor in the air, and on the shape of the crystal surface. For example, corners are more efficient than edges in attracting water molecules. For this reason the rods or radii of snowflakes always grow out of the six corners of the central platelet. As the flake falls through varying regions of a cloud, it encounters a variety of meterorological conditions. Since each flake follows a different path, this accounts for all the different shapes. However, at any one moment in its fall, the conditions immediately surrounding the flake are more or less uniform, so that all the exposed ends of the budding crystal are subject to identical influences. Presumably, then, they respond in identical ways, and hence the crystal preserves its symmetrical appearance as it grows.

The modern answer to Kepler's question involves three elements: atomism, an external agent in the form of the environment, and the idea of symmetry. The problem has occupied entire careers and profited from all the experimental and theoretical tools of physics. In turn, the study of crystal structure that began with Kepler's musings about close packing of cannonballs led to a lasting preoccupation of atomic physics with symmetry.

Symmetry is a powerful conception, but it can be deceptive. J. R. Newman put it well:

> Symmetry establishes a ridiculous and wonderful cousinship between objects, phenomena and theories outwardly unrelated: terrestrial magnetism, women's veils, polarized light, natural selection, the theory of groups, invariants and transformations, the work habits of bees in the hive, the structure of space, vase designs, quantum physics, scarabs, flower petals, X-ray interference patterns, cell divisions in sea urchins, equilibrium positions of crystals, Romanesque cathedrals, snowflakes, music and the theory of relativity.

The difficulty lies in sorting the wonderful from the ridiculous. Kepler escaped the trap set by the seductive and superficial similarity between honeycombs and snowflakes, even without knowing that the first owe their shape to the constraints of space while the latter are determined by the accidental structure of water molecules. Kepler was right in juxtaposing the two, because both illustrate a common abstract idea—the idea of hexagonal area-filling shapes. Their cousinship is made explicit by their mathematical descriptions, which, being abstract, transcend their ridiculous differences. "A mathematician," says mathematician G. A. Hardy, "like a painter or poet, is a maker of patterns. If his patterns are more permanent than theirs, it is because they are made with ideas."

Symmetry, in its most general sense, is nothing but pattern formed by regular repetition. The bilateral symmetry of the human body, eternal in its fascinating appeal, refers to the duplication of the left side by the right. The rotational symmetry of a regular hexagon stems from the sixfold repetition of a side. The translational symmetry of a picket fence comes about by iteration of the design of a simple stake. Each of the examples of symmetrical structure in Newman's list involves the repetition of some basic shape or property.

In science, symmetry owes some of its usefulness to repetition. If a portion of a pattern has been observed and the symmetry is either known or suspected, the whole can be predicted. Inasmuch as prediction is an essential element of science, symmetry is a valuable instrument in the scientist's toolkit.

Just as drawings and photographs reveal spatial symmetry, lists of attributes or measurements, appropriately displayed, reveal abstract patterns. A familiar example is the calendar. Days, weeks, months, and years represent four nested cycles of which only the first and last have natural origins and hence exist independently of the calendar. Old-fashioned tear-off wall calendars, consisting of individual leaves for each day (which flutter to the ground one by one in old movies to symbolize the passing of time), do not display the pattern. A matrix with days as column headings and weeks as labels of rows, on the other hand, subdivides the year into lines of seven days and blocks of about thirty. The cycles become explicit and demonstrate the repetitiveness, the symmetry, of the whole. Days of the week that share the same name, and often similar activities in ordered lives are grouped adjacent to each other, even though their dates are seven units apart. By use of this scheme we can predict what we will do on the nineteenth of July, because the nineteenth is listed under "Tuesday" and on Tuesdays we always play bridge. The repetitive rhythm of the calendar imposes on the monotonous succession of days a reassuring heartbeat without which life would be almost unthinkable. How differently we would perceive time if calendars were traditionally displayed in spiral, or triangular, or serpentine formats.

A scientific device similar to the calendar is Mendeleev's

Periodic Table that adorns the walls of science classrooms. Elements are numbered successively by weight, but listed in such a way that those that share common properties are contiguous, just as Tuesdays are listed below each other in the calendar. The significance of the scheme was demonstrated spectacularly in 1881 when the element germanium was discovered, twenty years after its prediction. By examining the elements surrounding a gap in his table, Mendeleev had been able to describe the physical and chemical properties of the missing substance with impressive accuracy. Derided at first as empty numerology, the Periodic Table put chemistry on a systematic basis and later, through its peculiar patterns, led to an understanding of the constituents of the elements—the atoms.

In our time similar schemes of classification have been applied to the particles that are produced by nuclear accelerators. The labels of the rows and columns are as arbitrary and abstract as the names of days in the calendar and of families of elements in the Periodic Table. They include such traditional attributes as electric charge and atomic weight, but also new exotic properties with such names as "strangeness" and "charm."

The most successful of the classifications of elementary particles is called, in imitation of the Buddhist prescription for right living, the Eightfold Way. By judicious choice of properties, the hundreds of particles that have been identified in nuclear collisions are grouped into families of eight members each. When the particles are depicted as points on a graph, the families form a familiar pattern. Two central points are surrounded by six others arrayed symmetrically. The periphery of the array forms a regular hexagon. Thus symmetry creates a ridiculous but beautiful cousinship between elementary particles and snowflakes.

The power of symmetry in making predictions was tested soon after the discovery of the Eightfold Way. In several cases the experimentalists had assembled evidence for the existence of incomplete families—patterns with gaps. In each case the properties of the known members allowed the prediction of the characteristics of the missing members. Exactly as in the case of germanium, these particles were eventually discovered with attributes very close to what had been predicted.

The Eightfold Way, like the Periodic Table, led further. Although it is an abstract pattern, it suggested an underlying structure, just as the shape of snow crystals reveals something about the architecture of water molecules. The success of the Eightfold Way inevitably suggests the Keplerian questions: Why do elementary particles come in groups of eight? What accounts for the hexagonal patterns of the Eightfold Way? The modern atomist turns the question around and asks: What can the symmetry of the pattern tell us about the constituents of elementary particles? He is not looking for shapes, as Plato was, because the hexagons do not exist in real space. Instead, the question refers to the abstract space of labels of properties of particles. It was found that if the so-called elementary particles are composed of even more elementary constituents called quarks, with fractional electric charges and fractional strangeness, then three types would suffice to build the whole messy zoo. This was in 1964. Since then most physicists have become convinced that quarks must exist in some sense or other, but so far no one has found them in the laboratory. Through symmetry nature hints at her secrets, but she does not reveal them easily.

For more than two millennia physics has fed on two great themes: the physical materialism of Democritus and the mathematical idealism of his contemporary Plato. Democritus based his world picture on atoms, whereas Plato believed in the supremacy of geometrical forms. At the very point where the moist warm vapors of atomism meet the cold purity of geometry, ideas condense into matter and form snowflakes, minute macrocosms that reveal by their symmetry the shape of their molecular constituents.

Atoms

A drop of food coloring and a Mason jar can demonstrate as vividly as any fat philosophical monograph the relation between the two ways of looking on nature, the artistic-emotional and the scientific-rational. Fill the jar with cold water, place it against a white background in a well-lit spot, let it stand for a few moments to settle down, and add a drop of blue. Observe carefully.

Like a miniature bomb the inky bead smashes through the surface into the water. In a fraction of a second, a fraction of an inch from the top, its fall is checked abruptly and turned into a gradual descent. As the droplet proceeds, it remains connected to the surface by a neck of color. The head expands and flattens. Soon the drop with its broadening stem looks like an upside-down mushroom burrowing its way into the depths. The urgent billowing of the cloud, its dynamic downward thrust, and its stark contrast with the clear water bring to mind the unfolding power of an inverted atomic explosion,

Atoms

but only for an instant. New phenomena soon capture the imagination.

The heavy globule at the head spreads out until it splits. For a moment it assumes the appearance of a thigh bone with its characteristic double-ended joint. Quickly the two halves separate, leaving a connecting bridge that lags behind, still attached to the stem. The long neck now appears as a faintly colored hollow tube, pale near the surface and darker farther down. The leading edges, still falling, are deep blue, but they are slowing. From the two halves of the drop, which may in the meantime have become quarters, and from the connecting tissue that reaches back up to the surface, thin sheets of pigment, like dyed curtains, slowly emerge. Eddies in the water cause them to spiral around, intertwine, overlap. Images of the aurora borealis are evoked.

To enliven the spectacle, add a drop of red. The spreading

of this dye is distinctly different in character from that of the blue. In some places the two colors overlap, in others they intermingle—in both instances producing purple. The descending clouds take on the fantastic shapes of ghostly dancers swirling many-hued capes in stately slow motion. Where the drapery hits bottom, the color intensifies. In the middle of the jar it diffuses. The whole spectacle is a sensuous psychedelic dream.

A drop of green enriches the stew. The motion of the whole is generally downward, but as the color spreads, an occasional veil wafts aloft. What was once clear is after a few minutes a marbled purple soup.

Rotate or tilt the jar: The structure within stays steady and unmoved. A gross intrusion, such as stirring or shaking, is required to upset the persistent stability of the delicate, colorful formation. Only one agent will ruin it—the force against which there is no defense, the destroyer of all art and beauty: time. Time dissolves the illusion. In half an hour the Mason jar has seen all order return to chaos.

The curious observer, watching all this, asks Why? How does it work? What's happening? What's the explanation? Why does it look like this, and not some other way?

The most perplexing and wonderful part of the display in the jar is the mixing of dye with water—the phenomenon of diffusion. A drop of oil behaves very differently from food coloring because it doesn't dissolve in water. In the shops of art and science museums, among the polished crystal and the Escher prints, one can buy expensive toys that duplicate most of the spectacle in the Mason jar, but because they make use of immiscible fluids they can always be brought back to the starting point. They do not demonstrate the reckless, irreversible destruction of order that is called diffusion.

Thinking about diffusion is like thinking about infinity: It begins innocuously but leads quickly to confusion and puzzlement. The first difficulty concerns divisibility. The drop of dye is evidently broken up into pieces that intermingle with the water. As long as the pieces are of reasonable size, the process is no different from the mixing of oil and water. By repeated division the individual droplets become too small to be seen, with only their cumulative effect visible as a faint color. But how far can this go on? How finely can matter be subdivided?

Suppose that there is no limit to the minuteness that can be achieved. Suppose every drop, regardless of how small it is, can still be subdivided. A point is reached where the portion of dye-matter is so tiny that it approaches nothingness. Where is the dividing line between matter and void? Between something and nothing? Does it make sense to talk of a piece of matter that is so small that no microscope can spot it, no scale weigh it? Could there be a portion of void that has similar properties? Do matter and void blend into each other smoothly and continuously? Or, if they are distinct, how is the difference defined?

Related to the problem of divisibility is the question of penetrability. Suppose the water is a continuum, without gaps or holes of any kind. The food coloring is another continuum. Somehow the two manage, by diffusion, to end up occupying the same space at the same time—a configuration that is strictly forbidden to ordinary objects like stones, flowers, and human bodies. Even before this miracle of interpenetration happens, how does the drop sink through the water? How does anything sink through a liquid? Imagine a marble, instead of a drop of dye, falling into the jar. At the front or leading surface of the marble, glass meets water, and if the marble is to move, the water must make way. Where shall it escape to? It can move around behind the marble, but only *after* the marble has moved forward to leave a space. Who shall make the first move—Gaston or Alphonse? Or shall they move simultaneously? How do front and back communicate to achieve this feat of synchronization?

The confusion is cleared up at one stroke by the atomic hypothesis that all material objects consist of tiny indivisible pieces of matter moving in a vacuum. The word *atom* comes from the Greek *a-tomos,* meaning "un-cuttable." (*Tomos* also finds its way into English via *tomography,* meaning the x-ray-ing of the body in slices, often with computer assistance, whence CAT scan.) Atoms solve the problem of divisibility by definition: No, declares the atomist, matter is not infinitely divisible. The vacuum solves the problem of penetrability by providing little empty spaces for the convenience of the front of the intruding object. The mystery of diffusion disappears. Molecules behave like ordinary macroscopic objects in that no two ever occupy the same spot at the same time.

What is surprising about the atomic hypothesis is not its uniqueness. All the questions that arise from diffusion could perhaps be answered in the context of a continuum theory of matter, without the introduction of atoms. But such an exercise would strain the rational faculties, whereas atomism is so simple, so astonishingly simple, that the modern reader, brought up with it since childhood, has trouble seeing why the questions of divisibility and penetrability pose any problems in the first place.

Armed with the atomic hypothesis, the watcher of the jar imagines the unfolding of a ballet. The dancers are molecules, small clumps of atoms permanently stuck together. As long as there are no chemical reactions in the water, each molecule retains its integrity. Water consists of an immense collection of identical molecules embedded in a vacuum and constantly on the move. They are very close together and slide around, with little interstices of vacuum to lubricate the motion. Food-coloring molecules are bigger and heavier than water molecules, and they respond differently to light, but their motions are similar. At the interface between dye and water the interaction between different types of molecules causes the surface tension that tends to keep the drops together and results in the formation of the veils. But surface tension isn't very strong. One after another the big blue intruders, jostled by some collision with their neighbors, shoot out of the dye droplet into the water, tumbling and bouncing all the way. In the water they meet a multitude of other foreign molecules that are invisible to the eye, impurities such as dirt, chemicals, air, and rust. Occasionally a large object floats by, composed itself of many strange molecules, a particle of sand, perhaps, or even a microscopic piece of vegetable or animal matter. Most of the time, however, the food-coloring molecule meets water molecules and vacuum.

The atomic hypothesis, the description of the universe as a vast collection of invisible material particles, is one of the oldest themes in natural philosophy. It arose, even before Aristotle's systematic examination of the world, from speculations about being and nothingness. The atom so conceived, the ultimate building block of matter, might be called the "philosophical atom" to distinguish it from later versions.

The most eloquent ancient exposition of atomism was written not by a Greek philosopher, but by a Roman poet, Titus Lucretius Carus. His poem *De Rerum Natura*, translated as *On the Nature of Things* or *On the Nature of the Universe*, appeared fifty-five years before the birth of Christ.

The poem is magnificent. In seventy-five hundred lines of hexameter in the heroic style, Lucretius ranged over a multitude of subjects such as matter, space, life, mind, sensation, cosmology, sociology, meteorology, and geology. Every phenomenon was ultimately reduced to the underlying basic atomic assumption that the poet took from earlier philosophers. The very names of his topics suggest that many of the explanations are neither modern nor particularly convincing. The problems of life, mind, sensation, and sociology are so complicated that atomism is not likely to be very helpful in their solution. An understanding of the constituents, which is the essence of atomism, is not always useful for the unraveling of collective phenomena; thus an analysis of the paint on the fenders of cars is irrelevant to a study of traffic patterns. But in the study of cosmology, meteorology, and geology, atomism has proved to be a good approach. In the study of matter it has been spectacularly successful from the time of Lucretius until today. Why this should be so is every bit as mysterious as why mathematics should work so well in describing nature.

Atoms for Lucretius are invisible, hard, imperishable bits

of matter that cannot be subdivided and that have no parts. They can bounce off each other and coalesce to form compounds. They are surrounded by vacuum in which they travel unimpeded. Different substances consist of different atoms, but the atoms of one substance are identical. In all these particulars the philosophical atoms of Lucretius anticipate the modern conception of atoms.

Historians warn of distortions caused by reading modern ideas into ancient texts. While this is good advice, the opposite is also true. Important insights, just because they were made long ago in a different age, should not be overlooked. It would be wrong to use the many wild and fanciful explanations of Lucretius to deprecate his grand scheme, which is correct in the light of modern science.

One of the principal techniques of ancient writers on natural philosophy, before experimentation became the touchstone of scientific truth, was analogy. To those who object to atoms on the grounds of invisibility, Lucretius answered:

Perhaps, however, you are becoming mistrustful of my words, because these atoms of mine are not visible to the eye. Consider, therefore, this further evidence of bodies whose existence you must acknowledge though they cannot be seen. First, wind, when its force is roused, whips up waves, founders tall ships, and scatters cloud-rack. Sometimes scouring plains with hurricane force it strews them with huge trees and batters mountain peaks with blasts that hew down forests. Such is wind in its fury, when it whoops aloud with a mad menace in its shouting. Without question, therefore, there must be invisible particles of wind which sweep sea and land and the clouds in the sky, swooping upon them and whirling them along in a headlong hurricane. In the way they flow and the havoc they spread they are no different from a torrential flood of water when it rushes down in a sudden spate from the mountain heights, swollen by heavy rains, and heaps together wreckage from the forest and entire trees. Soft though it is by nature, the sudden shock of oncoming water is more than even stout bridges can withstand, so furious is the force with which the turbid, storm-flushed torrent surges against their piers. With a mighty roar it lays them low, rolling huge rocks under its waves and brushing aside every obstacle from its course. Such, therefore, must be the movement of blasts of wind also. When they have come surging along some course like a rushing river, they push obstacles before them and buffet them with repeated blows; and sometimes, eddying round and round, they snatch them up and carry them along in a swiftly circling vortex. Here then is proof upon proof that winds have invisible bodies, since in their actions and behaviour they are found to rival great rivers, whose bodies are plain to see.

The power of Lucretius' song as a scientific text derives from its monolithic structure. Like Euclid, he set out his basic principles and then, to the end of the book, drew deductions from them with relentless vigor. This single-minded approach often led him astray, but the number of correct conclusions about physical science that follow from atomism is impressive.

For all that, Lucretius did not intend to write a scientific text; his purpose was humanitarian. The materialistic atomism of Lucretius is an alternative to the belief of his contemporaries in pagan gods. The first principle, the starting point, is the proposition that "nothing can ever be created by divine power out of nothing." The whole world, including human minds and souls, is a dance of atoms, untouched by interference from the gods. Everything we sense has a physical cause and can be explained without the hypothesis of divine intervention.

The motivation for Lucretius' heroic effort to convert his readers to materialism was his desire to release them from fear and terror of the gods, who apply their vast power capriciously and cruelly. The poem begins with the ominous words "When human life lay grovelling in all men's sight, crushed to the earth under the dead weight of superstition whose grim features loured menacingly upon mortals. . . ." For "superstition" many translations put "religion," because the Latin is *religio*. The crucial event that served to exemplify the horrors of *religio* was Agamemnon's sacrifice of his daughter Iphigeneia for the purpose of securing an auspicious wind for the fleet. Lucretius ended his characteristically graphic description of the hideous scene with the protest "Such are the heights of wickedness to which men are driven by superstition."

Even those who did not commit filicide were haunted by fears of eternal torment after death, and by terror of lightning, thunder, and other signs of divine wrath. It was to liberate them that Lucretius undertook the Herculean task of composing his poem.

He has been maligned and damned as an atheist by pagans and Christians alike. But while he denied direct divine intervention, his spirit was more reverential and moral than that of many of his professed religious critics. To Lucretius, nature, in its immense richness and subtlety, its manifold interdependences, its uniform laws and individual diversities, its fecundity and lavish generosity, its sheer beauty and wonder, was a worthier object of devout contemplation than a god who murdered animals and children. "True piety," he wrote, "lies in the power to contemplate the universe with a quiet mind."

The Epicurean materialistic philosophy of Lucretius did not survive the Roman Empire, but his atomism proved to be more robust. It disappeared from view, then surfaced again late in the Renaissance at the very moment when modern science was born. Two elements converged dramatically at that time. To the technique of rigorous mathematical deduction, perfected by the Greek geometers, was added the experimental method. In a short time around the watershed year of 1600 the discovery of how to discover was made simultaneously by a number of philosophers in Western Europe. In Italy, Galileo was laying the foundations of mechanics, itself the basis of physics, by rolling marbles down inclined planes and measuring the periods of pendulums with water clocks. In Germany, Johannes Kepler insisted on fitting the mathematical description to the observational data, and thus liberated astronomy from its idealistic preoccupation with perfect circles. In England, William Gilbert published his experiments on magnetism in a monograph that was to become a prototype for modern scientific textbooks. Also straddling the year 1600 was the career of the forgotten English polymath and atomist Thomas Harriot.

Harriot, who lived from 1560 until 1621, links the old science with the new. He also represents a link between the old world and the new, because he was the first English-speaking scientist in America. As a member of Sir Walter Raleigh's colony, Harriot spent the year 1585, more than two decades before the famous Jamestown colony was founded, on Roanoke Island in what is now known as the Outer Banks of North Carolina. There he studied the land, the people, the animals and plants of the region. He returned to England and published a wondrous little book called *A Brief and True Report of the New Found Land of Virginia.* It is really the first American book in English, and recommended reading for all who seek to discover their cultural roots in this country. Its long Elizabethan title ends with the words: "By Thomas Hariot; servant to the abovenamed Sir Walter, a member of the Colony, and there imployed in discovering. Imprinted at London 1588." What a phrase—"there imployed in discovering." A fitting motto for all scientists.

Thomas Harriot was a very great scientist indeed. Because his papers are finally being edited, the next fifty years will see his emergence from obscurity. He is mentioned occasionally in histories of mathematics, because some of his notes on algebra were published posthumously, but his best work concerns astronomy and physics. In July 1609 he used a telescope to look at the moon. Even nineteenth-century astronomers admit that "it is perfectly clear that Harriot and his friends had been in the habit of using telescopes before the discoveries of Galileo were known to them." His diagrams of sunspots, dated December 1610, are the earliest known observations. Three years later Galileo caused great controversy with his own publication on the phenomenon because it showed, in effect, blemishes on the most perfect of divine creations, the sun. Harriot observed the moons of Jupiter and calculated their periods, as his manuscripts clearly show. Again he anticipated Galileo. Why he didn't broadcast this spectacular discovery more vigorously, even among his friends, is difficult to understand. Harriot's observations of the comet of 1607, later to be known as Halley's comet, were good enough to be used in orbital calculations two hundred years later.

The fact that Harriot made all these observations was known to later astronomers. Around the year 1800 a German, Baron von Zach, attempted to rescue Harriot from obscurity but he was so severely rebuffed by his English colleagues, who apparently worshipped Galileo, that after a lifetime of struggle he had to give up in defeat. Harriot's papers were rejected for publication by his own university and both he and von Zach again disappeared from view.

In physics, Harriot anticipated some of the most important discoveries of the seventeenth century. After years of careful observation he formulated the law of refraction, or bending of light, that was found twenty years later by Willebrord Snell and bears the latter's name. Before Descartes he solved the beautiful problem of the radius of the rainbow. In 1605 he found that green and red rays of light bend in glass by different amounts, thus anticipating Isaac Newton by sixty years. He knew the differences between hexagonal and cubic close packing and in general had the insights into crystalline order that are usually attributed to Kepler.

Underlying all of Harriot's scientific work was a belief in atomism that set him apart from his more famous contemporaries. When Kepler asked for advice on some questions in optics, Harriot described to him how light entering a transparent medium travels through the vacuum, bouncing from atom to atom. The letter ends with endearing whimsy in a passage that serves even today as an enticing invitation to the study of atomic physics: "I have now led you to the doors of nature's house, wherein lies its mystery. If you cannot enter because the doors are too narrow, then abstract and contract yourself into an atom, and you will enter easily. And when you later come out again, tell me what wonders you saw."

Kepler did not follow Harriot's advice. He continued in his scholastic philosophizing about the union of opposites—transparence and opacity—and rejected atoms and vacuums. Harriot's theory of matter, based on Democritus, Epicurus, and Lucretius, involving eternal atoms continuously in motion, was truly a mechanical theory that served admirably not only in optics but also in the pioneering work on crystal structure. It linked up in Harriot's mind with a theory of mathematical atoms that was demanded by an analysis of infinite series. The periphery of a circle, for example, must be composed of an infinite number of atoms, or else it would not be possible to draw an infinite number of lines from the center to the circumference. Thus atoms underlie both mathematics and physics.

The great mystery behind the career of Thomas Harriot is Why is he unknown today? During his lifetime he was admired by many of those who had heard of him. But there were enemies as well. Harriot saw too far and too early—the court and the church could not stand for that. In 1591 Harriot was denounced as an atheist and conjurer. In the beginning of the seventeenth century, and indeed as late as the end of the nineteenth, atomism was linked with the doctrine of Lucretius and his atheism, and therefore considered heretical. Thus Lucretius, the poet who had sought to liberate humanity from the yoke of oppression by superstition, succeeded instead in impeding for two thousand years the easy acceptance of the theory of matter that he so passionately defended.

The history of atomism after Harriot is that of the evolution of a metaphor into a description of reality. As it unfolded, the philosophical atom of Lucretius was joined by different conceptions. The chemical atom, for example, began with the elements of Plato, acquired elaboration from the alchemists, and took its modern form in the early nineteenth century when it was considered a useful model, devoid of palpable reality. At the same time the theory of gases was developed into a sophisticated and successful mathematical structure based on the assumption of little hard spheres, the kinetic atoms. Chemical and kinetic atoms were considered by many to be no more than convenient artifacts.

In the first decade of this century the various models finally merged into the currently accepted real physical atom. Almost before it was born, however, the atom was undone. It was pried open and cut apart. Lucretius, whose atoms by definition were uncuttable, seemed discredited. But in the end he was vindicated when atomism moved on to a deeper level. (In a similar way symmetry under rotation is restored to the solar system when the focus of attention shifts from the elliptical orbits to the underlying equation of motion.) The physical atom was found to consist of more elementary objects—protons, neutrons, and electrons—which were prematurely called "elementary particles." These in turn were subjected to intense bombardment by nuclear accelerators in an effort to break them up further. For electrons, the attempt has failed: Electrons are truly uncuttable atoms of electricity. Protons and neutrons, on the other hand, are believed to be complicated little systems of the basic building blocks called quarks.

Quarks, like atoms, began as convenient fictions that allowed the consolidation of a large collection of experimental data into a simple pattern. These "mathematical quarks" are related to the classification of particles in the same way that the chemical atom was related to Mendeleev's Periodic Table. A different image that might be called "kinetic quark" emerged from the bombardment experiments. The manner in which various projectiles ricochet off a proton reveals the presence of tiny hard and heavy nuggets of some kind inside the target particle. The higher the energy of the projectile, the finer is the detail with which the interior of the proton can be mapped out. There is no evidence, even at the highest energies and the best resolution, that the kernels have any size. They are points and therefore fundamental in a way that protons never were. Their lack of structure lends tentative hope to the proposition that they may escape the fate of being subdivided in the future.

When mathematical quarks are identified with those hard, structureless pits or kinetic quarks, they take on a more tangible reality and become "physical quarks." They are today believed to be the building blocks of the material world.

Atomism, which has led from the philosophical atom of Lucretius to the modern physical atom and thence to quarks, is a powerful theme. Richard Feynman, Nobel Laureate and one of the architects of the description of matter in terms of quarks, a man admired by his colleagues for his complete and sure grasp of the whole physical world picture, put it this way:

> If, in some cataclysm, all of scientific knowledge were to be destroyed, and only one sentence passed on to the next generations of creatures, what statement would contain the most information in the fewest words? I believe it is the atomic hypothesis (or the atomic fact, or whatever you wish to call it) that *all things are made of atoms— little particles that move around in perpetual motion, attracting each other when they are a little distance apart, but repelling upon being squeezed into one another.* In that one sentence . . . there is an enormous amount of information about the world, if just a little imagination and thinking are applied.

Imagination and thinking are the keys that unlock the mysteries of science. With their aid the invisible particles of wind are made palpable, the narrow doors of nature's house open up, and a drop of food coloring in a jar permits a glimpse inside.

Warmth

Helen Keller's autobiography contains a celebrated passage in which the blind, deaf, and mute girl first discovers the meaning of language. It happens one day when her teacher decides to take her outdoors:

> She brought me my hat, and I knew I was going out into the warm sunshine. This thought, if a wordless sensation may be called a thought, made me hop and skip with pleasure.
>
> We walked down the path to the well-house, attracted by the fragrance of the honeysuckle with which it was covered. Someone was drawing water and my teacher placed my hand under the spout. As the cool stream gushed over my hand she spelled the word *water*, first slowly, then rapidly. I stood still, my whole attention fixed upon the motion of her fingers. Suddenly I felt a misty consciousness as of something forgotten—a thrill of re-

turning thought; and somehow the mystery of language was revealed to me. I knew then that W-A-T-E-R meant that wonderful cool something that was flowing over my hand. That living word awakened my soul, gave it light, hope, joy, set it free! There were barriers still, it is true, but barriers that in time could be swept away.

I left the well-house eager to learn. Everything had a name, and each name gave birth to a new thought. As we returned into the house, every object which I touched seemed to quiver with life. That was because I saw everything with the strange, new sight that had come to me.

This lyrical description of the learning process, with its haunting references to sunshine, light, and sight by someone who would never experience them directly, inadvertently singles out another sensation as fundamental, as prior somehow to consciousness itself—the feelings of warmth and coolness. The warm sun on the skin and the cool rush of water over the hand—these are wordless pleasures that we who can see and hear often take for granted, but that make deep impressions on those who are deprived of other, more overpowering senses.

The feeling of warmth is a curious experience, difficult to describe in words, and in some significant ways different from other sensations. The body detects light, sound, odor, and taste by specialized organs that respond to certain external stimuli. Similarly, shape, texture, and temperature of objects are recognized by means of the sense of touch that is located on the outer surface of the skin. In this respect, warmth is a stimulus to the senses, like light and sound. But warmth can also be felt in a general, diffuse sort of way throughout the entire body. Its origin can even be internal instead of external: Liquor and wine invigorate and warm as well as the sun does. The whole body, including its blind and deaf interior, can participate in the detection of warmth. Thus the feeling of warmth goes beyond what is usually meant by touch, and should perhaps be added to the list of the classical five senses.

It is significant that the detection of warmth is referred

to as "feeling." We "feel" warm and cold, and in that word there lurks a telling ambiguity, for feeling has two very different connotations. It means "exploring by touch," and in that sense it is appropriate for the perception of warmth by the skin. But feeling also means "experiencing an emotion," and this meaning is closer to what general, overall warmth causes in the body. Sociobiologists might conjecture that because warmth is required by a fetus long before it registers any other sensory stimuli, the sensation of warmth is more deeply etched in the psyche and represents a transition between the perceptions of the body and those of the spirit.

Be that as it may, warmth is required not only by the embryo from the moment of conception but, on a grander scale, by life itself. All animals need warmth for survival, even though in some cases they do not need very much. Furthermore, a current theory of the origin of life on earth holds that once upon a time, three or four billion years ago, lightning struck a puddle of water containing dissolved chemicals in a rich mixture resembling warm chicken broth, and triggered the formation of the amino acids that are the building blocks of life. Warmth played the same role in that primal scene that it plays today in the womb. It provides sustenance for life, and its absence is tantamount to death.

It was the control over this strange but indispensable commodity that signaled the emergence of man from the animals. When our ancestors learned to manipulate fire, they crossed the threshold to civilization. When warmth can be produced at will, during the night and in winter when the sun is ineffectual, intelligent control over the environment begins. In today's electrical world it is possible to produce synthetic cold as well as artificial warmth, but that particular accomplishment has played no significant role in the ascent of man. The gods conferred power on men by giving them fire, not ice.

Warmth shares with gravity the quality of ubiquity. Gravity, which permeates everything, can neither be turned off nor diluted, and it never ceases to operate. Our bodies, our artifacts, our very lives are built around the fixed fact of gravity. Warmth, although it can be controlled more readily than gravity, cannot be entirely eliminated either. Surreptitiously, inexorably it creeps back into places from which it has been banished. Everything, from the center of the earth to the outermost fringes of the universe, harbors some degree of warmth.

There is a big difference, however, between gravity and warmth. While the former displays very little variation in time and space, the latter is changeable in the extreme. Warmth comes and goes, from body to body, from room to room, from house to house, from hour to hour and day to day in ceaseless fluctuation. Its necessity for life and comfort, together with its fickleness, makes warmth a favorite subject for conversation. The temperature of the air is probably the most widely discussed, albeit scarcely understood, physical quantity. From the time we get up in the morning and choose clothes for the day, to the moment we go to bed and adjust the covers, we are concerned about warmth. All day long we wrap and unwrap ourselves, open and close windows, fiddle with heaters and air conditioners, warm food, cool drinks, turn hot and cold water faucets, and compare weather forecasts. We stamp our feet, rub our hands, fan our faces—and generally try to change the warmth that nature offers us by all means at our disposal. Unlike gravity, which we take for granted, we talk about warmth in minute detail. We worry when it is wrong, rejoice when it is right, and share our concern in endless redundancy with all who will listen. The ambient warmth is among the most reliable staples of human communication.

This prominent role of warmth in life, both as a physical condition and as a feeling, has influenced our language. The temperate adjectives *warm* and *cool,* as well as their more extreme forms *hot* and *cold,* share their literal meanings with a rich variety of figurative uses. A warmhearted person, a heartwarming experience, a warm friendship, a warm embrace, a warm disposition, a warm personality, a warm feeling—these are ardent, tender, lively, glowing, and very appealing. Cool, while it can signify deficiency in ardor, absence of enthusiasm, lack of interest, or worst of all, lukewarmth, is also used in a more positive sense. Keeping a cool head implies steadiness and control over passion. Heat, on the other hand, is often associated with unpleasantness because it exceeds the moderation of warmth. A hot temper, a hothead, a hot argument, to be hot under the collar—these are undesirable aberrations. Deviation in the opposite direction, a cold person, a cold shoulder, a cold heart, a cold stare, and a cold reception are just as objectionable. Mind and body favor moderation in matters of warmth.

Either by intuition, or by learning, everyone knows the comforts of warmth, the soothing pleasure of coolness, the misery of heat, and the piercing pain of extreme cold. But not everyone knows what warmth *is*. What is this stuff that causes our pores to sweat and our skin to glow red, whose absence makes our teeth chatter, our body shiver, and our skin pucker? What agent melts ice and turns water to steam? What cooks meat? What enters a match to make it burst out in flames? What invisible ingredient in the air lets flowers unfold and birds sing on a balmy spring day?

Philosophical speculation about the physical nature of warmth becomes scientific theory when measurements are introduced to turn qualities into quantities. The thermometer, developed in the seventeenth century on the basis of the ancient observation that warmth causes the expansion of materials, opened the problem to experimentation and brought precision and reproducibility. As is usually the case in science after a question has been raised, concepts have to be sharpened and terms defined before serious study can begin. Three terms that mean roughly the same thing in common parlance have to be sorted out. Temperature, heat, and warmth are used almost interchangeably but scientists look at them very differently. *Warmth* is regarded as merely a gentle form of heat and is rejected from the scientific vocabulary. It is heat in small quantities, and therefore not worthy of a special term. This leaves only *temperature* and *heat,* of which the first, as measured by a thermometer, refers to the intensity or strength of the phenomenon, whereas the second, less directly observable, is a measure of its amount or quantity. If warmth were a piece of cheddar cheese, temperature would measure its sharpness and heat its bulk. Temperature, like sharpness, can be determined from even a small sample of the material, while heat, like bulk, depends on the total amount of matter under consideration.

The word *temperature* is rooted in the verb *to temper,* meaning "to moderate." Temperature was originally the act of tempering, and acquired its current definition only with the introduction of thermometers. Thus moderation is at the very heart of the study of warmth. It would make more sense if physicists chose, as a measure of amount or quantity, the moderate word *warmth* over the extreme term *heat.* Indeed, in German the technical term is *Wärme,* while *Hitze,* which means heat, has no scientific standing. Thermodynamics is *Wärmelehre,* or "theory of warmth," a usage far more lavishly endowed with appealing connotations than our "theory of heat."

Besides the separation of the concepts of temperature and heat, another distinction has to be made before the true nature of warmth is understood. Fires, electric heaters, the sun, and all other luminous bodies emit radiant heat, which, upon strik-

ing the skin, produces the pleasurable sensation of warmth felt by Helen Keller. This kind of warmth is different from what is contained and stored in a warm object. Radiant heat shares many characteristics with light: It is stopped easily by the interposition of opaque screens, reflected by mirrors such as the shiny metal panels behind the elements of electric heaters, focused by lenses or "burning glasses," and transmitted through the vacuum of interplanetary space. The close analogy between light and radiant heat suggests the conjecture that the two are manifestations of the same basic phenomenon.

That radiant heat is indeed a form of light was proved serendipitously in the year 1800. Endeavoring to study the relationship between color and temperature, Sir William Herschel, the discoverer of the planet Uranus, produced a large color spectrum by directing sunlight through a prism. Into the various colors projected on a plain surface he placed thermometers—one for red, one for green, one for blue, and so on. As controls to establish the ambient temperature of the room in the vicinity of his apparatus he mounted additional thermometers below the red and above the blue edges of the artificial rainbow. To his astonishment the thermometer in the dark region beyond the red registered consistently higher than all the others. Herschel attributed the unexpected phenomenon to the presence in sunlight of an invisible component, called infrared light or radiant heat, whose rays are bent by a prism in exactly the same way that visible colors are. All warm objects, even if they don't glow visibly, emit radiant heat, which bathes the world in unseen luminosity. Owls, rattlesnakes, foot-soldiers with electronically enhanced vision, and spy satellites see in the dark by the illumination of radiant heat.

The fact that warm matter pours out infrared radiation does not imply that warmth is stored in the form of radiant heat. No more is it true that the light emitted by a candle is originally stored in the wax, or that the sound of a bell was once buried in the metal. Stored heat is, in fact, entirely different from radiant heat. This distinction, like the differentiation

Warmth

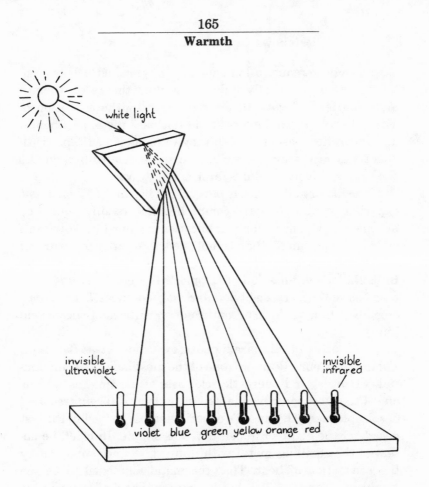

white light

invisible
ultraviolet

invisible
infrared

violet blue green yellow orange red

between heat and temperature, worried natural philosophers during the eighteenth and nineteenth centuries when the theory of warmth was developed. Once it is made, and the nature of radiant heat understood, the answer to the question, What is warmth? is unfortunately no closer.

The success of the fluid theory in explaining electricity prompted imitation. During Franklin's time, warmth was thought of as an invisible fluid, a material substance that flows from object to object like electric charge. But even analogy, that powerful theoretical tool, has its limitations. (A cautionary example is set by Lucretius who was led into error by enthusiasm for the success of his atomic theory of matter:

His attempt to reduce all phenomena, physical as well as mental, to atoms eventually failed.) The fluid theory of warmth and a variety of wave theories inspired by the confusion of stored heat with radiant heat did not pass the experimental and theoretical tests to which they were subjected. The truth about warmth is even simpler and more beautiful than the conjectures involving fluids, atoms, and waves.

The history of science is punctuated by a handful of classic experiments that are easily comprehensible, readily reproducible, and convincing in their implications. Some, like Millikan's oil drop experiment that established the particle nature of electricity, represent real scientific innovations. Others, like Buijs Ballot's verification of Doppler's law on the Rhine Railroad, serve the purpose of popularization of ideas. The famous demonstration of the true nature of warmth is of the second kind.

The author of this paradigmatic experiment was Benjamin Thompson, Count Rumford, one of the most ingenious, influential, versatile, and energetic scientists of his time as well as one of the most controversial, unscrupulous, self-centered, and despicable men. Thompson's experiment, performed in the last decade of the eighteenth century, was simple. In brief, he observed that drilling metal with a dull bit is accompanied by the production of heat. That the metal happened to be the barrel of a cannon, and that Thompson was drilling it in his capacity as general, head of police, and chief military advisor to the Duke of Bavaria, even though he was American by birth and a British Tory by persuasion, only adds piquancy to the story. The phenomenon itself is commonplace—it can be demonstrated using equipment no more complicated than a wire coat hanger. If the hanger is held firmly in both hands and the wire vigorously bent back and forth so as to break it, it will soon become too hot to touch. Heating by external friction, as in Thompson's experiment, or internal friction, as in the coat hanger, is nothing new.

What was new about Count Rumford's experiment was the conclusion he drew from it. Observing that he could bore for hours with a horse-propelled drill, and that heat continued to be produced in sufficient quantity to boil water and with no sign of diminution, he argued that heat cannot be a fluid because a material substance will eventually run out. If heat were a fluid that was originally stored in the drill or in the cannon or in both, it would at some point become drained and heat would stop flowing. In Rumford's own words: "Anything which any *insulated* body, or system of bodies, can continue to furnish *without limitation,* cannot possibly be a *material substance;* and it appears to me to be extremely difficult, if not quite impossible, to form any distinct idea of any thing, capable of being excited and communicated in the manner the Heat was excited and communicated in these experiments, except it be MOTION."

There it is at last: Warmth is motion. The motion of the horses is communicated to the cannon in the form of heat, and the motion of our arms is transformed into internal motion of the coat hanger that manifests itself as heat. The idea is sharpened when combined with the atomic view of matter, which supplies the precise locus of the motion: Heat is the invisible random jiggling of invisible atoms and molecules that constitute material substances.

Just as the existence of atoms was originally inferred from indirect evidence, the explanation of heat rests on a variety of arguments of which Count Rumford's is only one. Each by itself is inconclusive, but together they make an overwhelmingly powerful case. As each piece of the puzzle falls into place, the theory of warmth as random motion gains strength. The cannon-boring experiment concerns solids, but heat contained in fluids is the same thing. Applied to gases, the theory makes a particularly simple and convincing prediction. Confined in a closed vessel, gas exerts a pressure that can be interpreted as the force of a multitude of gas atoms bouncing off the walls. When the gas is heated, say by a candle under the vessel, the atoms move faster, crash into the walls more violently than before, and consequently increase their pressure. The increase of pressure upon heating has been carefully measured and corresponds in every detail to the calculation based on the interpretation of warmth as random motion.

The triumph of the theory of warmth is its startling economy of concepts. Instead of introducing new invisible substances, or fanciful complicated hidden mechanisms, the theory combines two of the oldest and simplest ideas in all of physics—the atomic hypothesis and the concept of motion. Atoms were invented in antiquity by the Epicurean philosophers. Motion was identified as a central concern of mechanics by Aristotle, and refined into modern form by Galileo. Combined, they explain the palpably sensuous phenomenon of warmth— completely different in kind from both ingredients of the theory and seemingly quite unrelated. Such parsimony producing rich and unexpected results is characteristic of the best not only in science, but in the arts as well. It is reminiscent of Picasso's evocation of voluptuousness by means of one deft line of charcoal, Beethoven's announcement of a mighty symphony by means of a few impeccably placed notes, the uncluttered purity of the Acropolis, a line of Keats, a paragraph of Virginia Woolf. Clarity and simplicity of the underlying elements, elimination of extraneous conceptual baggage, condensation and crystallization are among the desiderata of all forms of human expression.

Thermodynamics is based on the division of the properties of atoms into two kinds. Intrinsic or fixed attributes include size, structure, shape, weight, electrical charge, and chemical affinity. They largely determine the intrinsic characteristics of substances, such as color, density, translucence, and so on. A different kind of property, of which the chief example is motion, is imposed on an atom by the outside world. The speed of an atom is not a fixed quantity, but can take any value from zero to the speed of light. Warmth too is a variable property. The same piece of steel, with little change in outward appearance, can have different temperatures. The similarity between the speed of microscopic entities and the warmth of macroscopic entities, that they are both variable and susceptible to external manipulation, is not very suggestive, but it sufficed for the inventors of the theory, like Count Rumford, to relate the two. This creative association of the two seemingly incongruous concepts of heat and motion is so rewarding, and its aesthetic appeal so beguiling, that it ranks with the supreme achievements of physics.

While the theory of relativity, another masterpiece of physics, rises like a sonata from certain subtle mathematical contradictions between mechanics and electromagnetism, the theory of warmth is at home, like a peasant folksong, in the kitchen. Stoves and chimneys, rather than calculus and wave equations; pots and pans, rather than interferometers and cathode ray tubes; maids and soldiers, rather than mathematicians and philosophers, inspired thermodynamics. Count Rumford, the Yankee adventurer, pursued his science in the same way that he led his life—pragmatically. Throughout his career he was preoccupied with the application of heat. Why does apple pie burn the mouth? What prevents a chimney from smoking? How can food be cooked with less fuel? Why do fur coats retain warmth? How can steam be kept in a pot? What foods provide the most nutrition for the least cost? What is the best way to roast mutton? What is the most efficient design for a portable coffee pot?

To many of these questions, Rumford, both as private scientist and as commander-in-chief of the army, found answers that survive to this day. His fundamental improvement of fireplaces is the basis for all modern designs. His woodstoves, more sophisticated than Franklin's, were based on theory as well as extensive experience, with the result that they performed better than many of today's reinventions. Ranges in large kitchens, before the advent of electricity, owed their layout to Rumford. The notion that it is the air trapped in furs that insulates the wearer from the cold is now accepted as correct. The little lip on the rim of expensive cooking pots that traps water for sealing the lid was perfected by Rumford.

Not all of Rumford's ideas were correct, of course. His thinking about heat was flawed by great misconceptions. One of the most fascinating is the notion of "frigorific rays," or radiant cold, in analogy to radiant heat. "Cold," he said, "can with no more propriety be considered as the absence of heat than a low . . . sound can be considered as the absence of a higher . . . note." The perils of misplaced analogy! Frigorific rays were discussed and apparently even demonstrated in Rumford's time. If they really existed they would be used today in fast-food restaurants. Hamburgers could be kept warm by infrared radiators on the counter, next to ice cream kept frozen by frigorific lamps of Rumford's design. To him, the usefulness of the theory of warmth was just as appealing as its intellectual power. "I can conceive of no delight," he wrote in 1797, "like that of detecting and calling forth into action the hidden powers of nature!—Of binding the Elements in chains, and delivering them over, the willing slaves of Man!"

After his death, as the study of warmth turned from teakettles to steam engines and from fireplaces to power plants, its theoretical foundations grew deeper and stronger. The fundamental principle that heat is motion was joined by a rule as profound in its implications as it is innocuous in its statement. Called the second law of thermodynamics, it states that "heat flows naturally from a warmer body to a cooler one." The empirical basis for this observation is so ordinary that it hardly bears mention. Cold water in a pot on a hot stove warms up, rather than losing the remainder of its heat to the stove and turning to ice. Whenever hot and cold objects are in contact, warmth flows from the higher temperature to the lower, regardless of which object is larger and which contains a greater total quantity of heat.

The principle underlying the second law of thermodynamics is a second surprise, as intriguing as the association of warmth with motion. The operative concepts are chance, disorder, randomness, and probability. The compelling rigor of the second law appears to be antithetical to such notions, but the contradiction is superficial. Probability is the study of laws governing random events. The toss of a coin, for example, is random, so that its outcome cannot be predicted, no matter how many tosses have gone before. The question "Which side will come up next?" cannot be answered with any confidence. If the problem is posed more cleverly, however, answers are to be found with increasing reliability as the number of tosses increases. In a string of ten, how many will come up heads? The answer is "Roughly half." The fraction of heads will be much closer to a half for a million throws than for ten. For the astronomical numbers involved in the counting of atoms and their configurations, probabilistic laws approach certainty.

Physics seeks to discover regularities in the behavior of matter. Atomism breaks matter down into innumerable particles, which, according to the theory of warmth, are in random motion. Statistics restores order at a different level in the hierarchy of analysis and makes possible the exact science of thermodynamics.

Specifically, statistics is applied to the distribution of positions and speeds, the fundamental mechanical properties of atoms and molecules. Randomness in these attributes constitutes the basic assumption of the theory of warmth. A drop of ink that falls into a bowl of water quickly diffuses throughout the container. The opposite scenario, a mixture of ink and water that spontaneously separates into clear water and a jet-black droplet of ink, is unheard of, not because it is impossible, but because it is unlikely. There are many more possible arrangements of molecules of ink when they are spread around the bowl than there are arrangements of the molecules in one tiny spot. Therefore, as the particles meander and take up different positions, it is overwhelmingly more probable that they find themselves in one of the numerous configurations of uniform distribution than in one of the few configurations of confinement to a drop. Disorder is more likely than order in the distribution of molecules of ink in a fishbowl, as it is in the distribution of toys and clothes around the room of a three-year-old.

The same reasoning, applied to the more abstract notion of speed, accounts for the second law of thermodynamics. If a hot pebble is dropped into a bowl of cold water, the rapid pebble molecules jostle the water molecules and pass on their motion. It is most probable that the speeds of all the molecules, in water and pebble alike, will end up being roughly equal. If all the possible divisions of a given quantity of motion among the molecules of the system were listed on some immense computer printout, the cases that correspond to roughly equal speeds for all particles would overwhelmingly outnumber those that are characterized by some special arrangements, such as high speeds concentrated in the pebble. Again, disorder is more likely than order.

Disorder in position accounts for diffusion, disorder in speed for the second law of thermodynamics. In both cases, the tendency of nature toward greater disorder imposes an arrow of time. If a movie of two billiard balls colliding on a pool table is shown forward and backward, it is impossible to tell the difference between the two. A movie of ink dropped into water, on the other hand, can be properly shown only in one direction. A hot pebble in a pot of water also displays an evolution of temperature that proceeds in a unique direction. Perhaps time, that ineffable quantity, can be defined as the parameter that measures how disorder, driven by the second law of thermodynamics, progresses.

Nevertheless, it must be remembered that diffusion and the flow of heat can be reversed. Ink and water can be separated, a pebble can be warmed up at the expense of the heat stored in the bowl, a messy room can be straightened out. However, each of these acts costs effort. By themselves they

don't happen. If left to its own devices, the world tends toward complete disorder and uniform lukewarmth. That is the significance of the word "naturally" in the original formulation of the second law of thermodynamics: "Heat flows naturally from a warmer body to a cooler one." Reversing the natural trend is possible but costs energy.

That the universe is becoming more disorganized, that its molecules spread out ever more uniformly, that the hot stars will cool off and the molten lava freeze, that all differentiations will ultimately cease, leaving the cosmos a smooth homogeneous lukewarm gruel—that is a melancholy thought. While warmth means life and joy and pleasure, and fire evokes images of kitchen, forge, and factory, the second law hovers in the background, promising an eventual cold death. The languid world-weary Victorians, exhausted by the exuberance of the Industrial Revolution that had been brought about by the harnessing of heat, were much distressed by this notion and worried about the ultimate fate of the universe in what was called "heat death," a sad ending and difficult to reconcile with God's bounteous generosity. Percy Bysshe Shelley captured the mood in "Ozymandias":

I met a traveller from an antique land
Who said: Two vast and trunkless legs of stone
Stand in the desert . . . Near them, on the sand,
Half sunk, a shattered visage lies, whose frown,
And wrinkled lip, and sneer of cold command,
Tell that its sculptor well those passions read
Which yet survive, stamped on these lifeless things,
The hand that mocked them, and the heart that fed:
And on the pedestal these words appear:
"My name is Ozymandias, king of kings:
Look on my works, ye Mighty, and despair!"
Nothing beside remains. Round the decay
Of that colossal wreck, boundless and bare
The lone and level sands stretch far away.

Today, we are less resigned. So much has been learned since the turn of the century, when physicists presumed to know all the laws of nature, that we are sure of nothing save our own ignorance. The world around us does indeed seem to run down according to thermodynamics, but who knows what lurks out there in the vastness of outer space? Current theoretical speculation suggests prodigious sources of energy, as yet undiscovered, that may fuel the effort required to create new order and to overcome the ominous decree of the second law. Matter in the vicinity of strange places like black holes probably behaves according to new and different, and perhaps more optimistic, laws. God, beginning with chaos, separated light from dark, and land from water. He also had to exert His vast powers to separate hot from cold. Perhaps He has unseen ways of imposing new order even now, and of restoring warmth to places that have cooled off.

The Measure
of Things

The 1979 Nobel Prize in physics was awarded to Sheldon Gla-
sow, Abdus Salam, and Steven Weinberg for contributions to
the theory of elementary particles. Those who keep score by
nation can chalk up two more for the United States and one
for Britain. The statisticians can analyze finer details such
as distribution by country of origin, country of residence, religious preference, sex, and color of hair (Salam is from Pakistan,
Weinberg a redhead). Such numbers are not only irrelevant,
but also distracting. They obscure the fact that in addition
to people, the Noble Prize honors ideas and thereby transcends
nationality, sex, creed, and color. Among the impressive intellectual oeuvre of the three recipients, one idea stands out.
Its name is even less familiar than words like *quark, charm,*
and *strangeness* that pepper the working vocabulary of particle
physicists. The main idea honored in 1979, called the "gauge
principle," is destined to outlive much of the current jargon.

By itself, it makes little sense, but like a figure in a mural it acquires meaning in relation to its background. The background consists not so much of a welter of observations and theories as of a small number of concepts at the basis of the physical world picture, themata or leitmotifs that determine not only the way in which answers are found, but the kinds of questions that are asked about nature.

The themes are simple. The really deep insights that lie at the center of the most far-reaching theories, especially in the physical sciences, are surprisingly uncomplicated. Unfortunately, the essence of physics is buried, like a pearl in an oyster, under deep soft layers of interpretation and jargon, which in turn are protected by a tough crust of mathematics. To most people the outer shell seems impenetrable and the viscera undigestible. Only by drawing word pictures, by making analogies, can a physicist reveal the pearl without embarking on the arduous job of making the oyster palatable. Like every translation, the representation of formulas by words is far from faithful, but because the underlying ideas are simple they survive the transformation. "Though analogy is often misleading," said Samuel Butler, "it is the least misleading thing we have."

The themes are also old. Radical new ways of looking at nature are extremely rare. Most ideas come and go throughout history, appear in different guises, take on different meanings, pop up unexpectedly in unfamiliar contexts, alternate between acceptance and rejection. Metamorphosis is more common than spontaneous generation not only among animals, but also among ideas.

One of the simplest and oldest recurring ideas in physics is that of atomism. The existence of atoms was passionately debated by Greek philosophers. As the centuries passed, acceptance of the atomic hypothesis rose and fell. By the beginning of the twentieth century, the reality of atoms had been established; they were pictured as miniature billiard balls in continuous motion. What had once been a philosophical hypothesis had become a fact, but with a crucial change. Experiments revealed that the new atoms were eminently cuttable. Their outer layers were stripped off, then the inner core was split, and its parts split again. Quarks, the final products of this frenzied smashing, can be detected only indirectly. Together with electrons they are thought to be the basic constituents of matter, the atoms of our age.

Another simple idea that is as old as natural philosophy itself is the concept of symmetry. Regular shapes are pleasing to eye and mind. It is a delightful surprise to make out, in the infinite random carpet of stars, the pattern of Orion's belt. It is magical to watch a growing crystal acquiring perfectly plane sides and square corners in a beaker of salt water. It is soothing to follow the circular ripples spreading under a gentle rain. Like painters and poets, physicists have searched for symmetry in nature from the very beginning. Aristotle taught that circular motion, being perfect, is divine—and vice versa. He insisted therefore, and for almost two thousand years his opinion prevailed, that all celestial motion must be circular. It was a terrible intellectual shock to discover, with Kepler, that planets trace out ellipses. Ovals are pleasing, too, but they are not as perfect as circles. The idea of divine symmetry seemed badly shaken. It soon reappeared, however, at a deeper level when Newton showed that perfect circular symmetry is exhibited not by the actual motion of the planets, but rather by the mathematical equation governing the motion. It is this formal property of Newton's second law, in combination with his Universal Law of Gravitation, that excites physicists. The symmetry of the equation is as obvious and as pleasing to the trained eye as the symmetry of a circle is to the untrained. The old Greek obsession with circles has been replaced by a mathematical theorem, but the circular orbit can still serve as a metaphor for the underlying mathematical symmetry.

A third ancient and obvious idea is relativity. The word now connotes two complete theories, the special and the general, both formulated by Albert Einstein, but the concept is much older. In its simplest sense, relativity refers to the realization that all change, and specifically motion, is relative.

A good example is the daily motion of the stars. The whole sky seems to revolve in stately procession around the earth every twenty-four hours. More than two thousand years ago Heracleides recognized that the same appearance can be explained by assuming a rotation of the earth on its own axis under a fixed firmament. In both descriptions of the universe, the relative motion of observer and stars is the same. Much later Galileo, in analyzing the parabolic flight of projectiles, discovered the principle of relativity that Einstein eventually used as the first of two axioms in the special theory of relativity. It states that the description of nature inside a reference frame, say the hold of a ship at sea, is exactly the same as it is when the ship is at rest in the harbor. The moving frame and the stationary frame are equivalent. An important consequence of this idea is the impossibility of finding an object that is truly at rest. If I declare that a billiard ball, say, is at rest on a table, an observer in a passing car is entitled to claim that the ball is really in motion. The principle of relativity therefore acts as a sort of enabling legislation for uniform motion of every object with mass in the world.

Atomism, symmetry, and relativity are among the powerful themata of physics. Now the "gauge principle," more complicated than the others but interwoven with them, has been added to the list. The *Oxford English Dictionary* puts the earliest use of the word "gauge" in the thirteenth century when it meant a fixed or standard measure. Later the meaning broadened to include devices for making measurements, such as tire and rain gauges. *Roget's Thesaurus* lists fifty-one measuring instruments under the heading "gauges." In the twentieth century, anglophone physicists adopted *gauge* as a translation of "Mass-stab," the German word for measuring rod.

Unfortunately, and somewhat surprisingly, etymology is of no use whatever in explaining the gauge principle. It is a mathematical property of certain theories, and has nothing to do with measurement. For our purposes, circumlocutions and similes have to take the place of a proper mathematical definition. Nevertheless, even a vague glimmer of understanding is worth striving for because the gauge principle is destined to assume a central role in the physics of the future. It is fitting that the name of such an important idea is etymologically linked to the concept of measurement because quantification and measurement are fundamental to science. In a very tenuous sort of way the phrase *gauge principle* may therefore be apt after all.

What physicists hope to achieve by means of the gauge principle is nothing less than the unification into one theory of all the forces of nature: gravity, electromagnetism, the "weak force" responsible for the radioactive decay of atoms and particles, and the "strong force" that holds nuclei together.

The unification of the description of various forces is one of the principal aims of modern physics. William Gilbert showed in 1600 that the magnetic influence of the earth is of the same nature as that of the lodestone in his laboratory. In 1666 Isaac Newton proved that the force that pulls an apple from its tree also keeps the moon in her distant orb, thus unifying the seemingly separate phenomena of terrestrial and celestial gravity. Benjamin Franklin unified atmospheric and artificial electricity in 1752. Hans Christian Oersted demonstrated in 1820 that an electrical current exerts the same force as a magnet. In the 1860s James Clerk Maxwell succeeded in showing that electricity and magnetism, phenomena as different as amber and lodestone that gave them their names, are manifestations of one single effect, called the electromagnetic force. Albert Einstein spent his latter years in an unsuccessful search for the unification of electromagnetism and gravity. The purpose in each case was economy of description: the parsimonious use of the smallest possible number of assumptions and concepts to describe the greatest possible variety of phenomena.

Much recent excitement in elementary particle physics stems from the experimental verification of a unified theory of electromagnetic and weak forces. Called somewhat unimaginatively "the electroweak theory," it is based on the gauge principle. A theory of the strong force, which is not yet as well established as the electroweak force but which has received considerable empirical corroboration, also builds on the gauge principle. The ultimate combination of all known forces is described with bombast by such phrases as *grand unification, supersymmetry,* and even *maximally supersymmetric grand unification.* The ultimate theory has not yet been developed, but it is generally agreed that the gauge principle will be an essential ingredient in its formulation.

But what is it, this "gauge principle"?

Imagine a blue marbled bowling ball resting in the middle of a smooth green cloth. Imagine also, some distance away on the same baize, a white billiard ball. An observer in a balloon high above the scene sees both balls, the latter appearing as a white dot. Watching closely she notices that the dot begins to move straight toward the center of the bowling ball, imperceptibly at first and then with gathering speed. If she is of a philosophical bent, the observer will wonder about the motion of the dot, and will try to interpret the phenomenon.

To simplify the question, instead of considering the full course of the motion, she focuses only on a brief interval of time during which a tiny distance is covered. She believes that if she can describe that small portion, the continuous, orderly, and causal nature of the world will allow her to predict the next step, and then the one after that, and so on to the end. Such an analysis amounts to a kind of atomism of motion, the hope that understanding of the parts leads to mastery of the whole. The Greeks, and later philosophers up to Galileo and Kepler in the seventeenth century, were concerned with the geometry of complete trajectories, the straight lines, circles, parabolas, ellipses, and hyperbolas found in nature. Newton's step-by-step method, referred to as calculus in mathematics, permits the derivation of these same shapes, but also of other figures of vastly greater complexity.

The woman in the balloon has three alternative explanations for why and how the billiard ball moves. They correspond to three theories of the nature of forces.

A simple, old, and truly magical analysis interprets the

white dot as a particle influenced in some way by the bowling ball. The influence—a pull or attractive force—reaches out radially from the bowling ball. Its strength and variation with distance depend on the properties of the two objects on the cloth, but its origin remains unexplained. Such an interpretation, called action-at-a-distance, is as mysterious to us as it was to Newton, who invented it.

The trouble with action-at-a-distance is that it violates intuition. Material objects, when not in contact, ought not to influence each other at all—unless there is some intervening medium or mechanism. Yet gravity, electricity, and magnetism were all described as action-at-a-distance, and for a long time the explanation sufficed.

The second explanation that might occur to the philosophical balloonist is more subtle. Rejecting action-at-a-distance,

Rainbows, Snowflakes, and Quarks

she assumes that the billiard ball responds only to its immediate environment. In particular, she might assume that the appearance of the cloth is deceptive. It is really distorted and looks flat only from above. If the baize is stretched like a drumskin over a frame without tabletop, it will be indented in the middle by the weight of the bowling ball. Everywhere it slopes gently toward the center. Any object placed on the cloth will begin to slide down the slope. The observer from above cannot see the warping of the cloth, but she can see the motion of objects. Furthermore, two objects placed close together experience the same incline and hence proceed to move with equal speeds, as required by Galileo's law that all masses fall with equal acceleration.

This picture illustrates the curved space of general relativity. There is no direct force between the two balls. Instead, space, represented by the cloth, is curved by the presence of the large mass and in turn affects the smaller one. Space mediates between the two objects. Cause and effect are proximate: An object is influenced only by the shape of the cloth that it touches. Unfortunately this analogy, though suggestive, is cir-

cular because it requires a downward gravitational pull on the billiard ball to explain the lateral motion. Since the model is supposed to *explain* gravity, that requirement is a flaw. A better simile eliminates the blemish but complicates the issue by explicitly introducing the notion of time. General relativity is a theory of gravity by means of curved space-time, an unimaginable conception indeed.

The geometrical explanation of gravity provided by general relativity, despite its conceptual difficulty, is completely successful. However, until now it has seemed impossible to incorporate other forces, such as electromagnetism and the strong or weak forces, into the same scheme.

The third alternative for the intrepid balloonist is an analogy of the gauge principle. She again imagines the cloth to be flat, but she rejects action-at-a-distance. As in the second image, the billiard ball is assumed to respond only to conditions in its immediate vicinity—it is influenced by direct, local interaction with the things it touches. Because the cloth, representing space, is level, it offers no clues for an explanation of the motion. Instead, the balloonist shifts her attention to the ball itself.

If the ball is perfectly round and smooth, with no markings or spots of color, a rotation on its own axis is undetectable. Spherical symmetry causes the ball to appear identical from all directions, and rotation to be invisible. Armed with the most powerful telescope, the balloonist cannot determine whether the ball is rotating or not. Translational motion of the center of the billiard ball, on the other hand, is easy to see and measure with reference to the stationary bowling ball.

The balloonist realizes that she should consider the possibility of rotation of the billiard ball, simply because she cannot rule it out. As soon as the billiard ball is assumed to rotate while in contact with the cloth, it will of course begin to roll. The explanation for the motion of the white spot in this crude analogy for the gauge principle is simply the rolling of the billiard ball. No external force is postulated, nor a curvature of space, only the possibility that a symmetrical ball may, in fact, be rotating.

For all their inadequacy, the models illustrate an important point concerning the kinds of forces that are capable of being incorporated into the theories. Action-at-a-distance suffers from an excess of generality. An infinite number of different forces can be described by the formalism. In the past, when it was used to analyze electricity, magnetism, gravity, and many other macroscopic phenomena, generality was a virtue because it allowed diverse effects to be described within a common framework. Today, however, the emphasis is on restricting the number of alternatives, on finding a formulation that includes only those forces that are found in nature, but forbids others. Action-at-a-distance is of no help in this endeavor. General relativity errs in the opposite direction. It is so limiting that only gravity, and no other force, fits within its narrow boundaries. In the model, for example, any object placed on the cloth experiences the same acceleration toward the center. Repulsions, such as the electrical force between two like charges, and such lateral forces as magnetism are excluded. The third model illustrates a middle ground between generality and limitation. By adjustment of the rotation of the billiard ball in a forward, backward, or sideways direction, attractive, repulsive, and even transverse forces are possible. Nevertheless, even though the magnitudes and directions are adjustable, only a small number of forces, those that result in rolling motion of a billiard ball fit into the picture. The gauge principle rules out an infinity of imaginable forces while at the same time permitting those that exist in nature.

The actual gauge principle differs from the model in a crucial respect. A fundamental particle, such as a quark or

an electron, does not have the three-dimensional form of a billiard ball. It is an a-tom in the original sense of un-cuttable: It possesses no parts, no structure, no size. It is a mathematical point, and points cannot rotate. Instead, it is assumed to be endowed with fictitious, internal axes—a set of imaginary lines, sticking out like radio antennas from a miniature spherical sputnik. There may be one axis, or two, or a large number, depending on the type of particle. Since the axes are hypothetical, the orientation of the whole imaginary structure cannot really matter; turning it leaves everything unchanged. The theory is said to be symmetric under rotations around the axes, in the same way that the billiard ball is symmetric under rotation. But symmetry under rotation *requires* the possibility of rotation, as the balloonist understood. (Recall how the Galilean principle of relativity *requires* the possibility of uniform motion in a straight line.) Now it happens that there is a connection between the fictitious internal axes and real three-dimensional space. This connection, a highly mathematical sort of link, is symbolized by the contact of the billiard ball with the cloth. The connection is such that as the particles rotate, it changes its motion through real space. The new theory thus offers an alternative to the usual description of acceleration in response to external forces. The gauge principle, in its strongest formulation, asserts that all forces in nature are due to this mechanism, and that they can be combined into one unified formulation by proper choice of the internal axes.

The introduction of a fictitious set of internal axes, followed by the requirement that rotation is undetectable, seems to be an unnecessary complication and a violation of the principle of parsimony—but the prescription obviates the need for introducing other ad hoc assumptions about forces and makes the price well worth paying. After working with theories based on the gauge principle for a while, physicists find the extra step as natural as switching mentally from sliding billiard balls to rolling ones. Once that essential step has been taken, the theory is so rich and yet so straightforward that it exerts

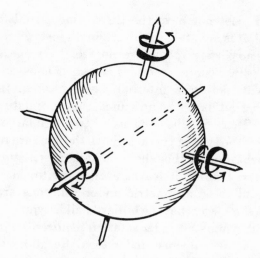

a seductive aesthetic appeal. What makes it a serious physical hypothesis is that it leads to very specific and detailed predictions about the properties and interactions of particles. To the extent that they are verified, the additional assumptions are justified.

Adoption of the gauge principle has allowed the unification of electromagnetic and weak forces as first suggested by Steven Weinberg, then of Harvard University, in 1967 and by Abdus Salam of Imperial College, London, in 1968. Strenuous experimental and theoretical elaboration of the idea has convinced the community of physicists that it is sufficiently plausible to deserve a seal of approval in the form of the Nobel Prize. The change inflicted on the electromagnetic force by its marriage to the weak force is very small, but it has been measured. It is not the tiny correction to Maxwell's theory, however, that is significant about the Weinberg–Salam model. Rather, it is the possibility of progress toward the unification of all known forces. It is a step in the direction of the goal pursued by Gilbert, Newton, Franklin, Oersted, Maxwell, and Einstein.

The "gauge principle" will take its place among such venerable themes as atomism, symmetry, and relativity as a fundamental insight into the workings of nature. If history is a guide, its formulation will become simpler and more comprehensible with time. New descriptions of its content and new analogies will be developed. Eventually it will become as common as atomism, as apparent as symmetry, and as comprehensible as relativity. In the meantime we can try to be content with the humble image of a smooth white billiard ball rolling along on a green baize cloth.

Applied to quarks, the gauge principle leads to a force with strange new properties. Gravity and electromagnetism, the familiar forces of macroscopic everyday life, diminish in strength with distance. If the earth, for example, were far enough removed from the sun, the mutual gravitational attraction of the two would become so weak as to be ineffectual. Conversely, as two bodies approach each other, these forces become stronger. Quarks interact gravitationally and electromagnetically, but in addition they have internal axes that correspond to a different and stronger force with a behavior that is exactly the opposite.

The "strong force" between quarks grows stronger with distance. Such perverse behavior was unexpected for particles but it is common in everyday life. Rubber bands and steel springs have just that property. The longer they are stretched, the more difficult it becomes to elongate them further. Quarks are believed to be bound together by a similar force based on the gauge principle. At close range it is so weak that the quarks, like marbles linked by slack rubber bands, hardly affect one another at all, but at distances of a few nuclear diameters the force becomes insurmountable. According to this theory, quarks are destined to remain near each other. They are said to be confined to the nucleus, and will never be found in isolation.

Atomism has taken a new turn. Constituents that can never be isolated are not, strictly speaking, constituents. The point of departure of the atomic doctrine is precisely the hope that the world can be seen as an edifice of elementary building blocks that can be separated, studied, manipulated, and put

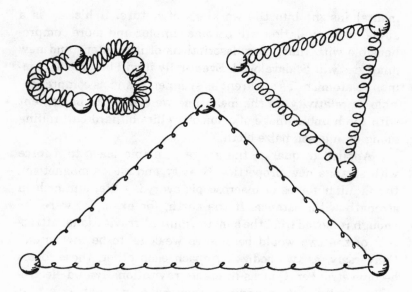

back together—but quarks don't answer to this description. They are building blocks that are forever tied together, a state of affairs that borders on the paradoxical. Nature delights in surprises. At the beginning of the century, physicists had to accept the notion of cuttable atoms. Later they found that particles behave sometimes like massive points, at other times like waves. In the second half of the century the nonelementarity of elementary particles surfaced again. Now it seems that the notion of particle has to be revised once more to incorporate the possibility of permanent confinement in more complex entities.

That a theme as ancient and basic as that of atomism undergoes modification is reassuring. Physicists are keenly, sometimes painfully, aware of the tentativeness of their conclusions. Experiments and theories undergo continual refinement in the interest of an increasingly faithful description of nature. The themata, however, the assumptions that underlie both observation and analysis, tend to remain unspoken and relatively unchanged. Their constancy lends historical unity to

The Measure of Things

the enterprise of physics, but carries with it the risk of stagnation.

Ultimately, however, even the simple, ancient themes come up against hard reality, and must yield if they are inadequate. Subtly they are modified, intuition is remodeled, concepts are merged into new combinations. Scientists struggle to fit their ideas into comfortable molds, but if nature decrees they don't fit, the molds have to be broken and replaced. Stubborn facts provide safe anchors against escape into a world of theories built on neat but erroneous preconceptions.

Not the scientist, but nature has the last word.

Bibliography

The following are my most important sources. All were written for non-specialists, and most provide references for further reading.

INTRODUCTION. The idea of themata, and much of the inspiration for my approach to physics, are to be found in the work of Gerald Holton, especially his *Thematic Origins of Scientific Thought—Kepler to Einstein* (Harvard University Press, 1973) and *The Scientific Imagination* (Cambridge University Press, 1978). The quotation from J. C. Maxwell that opens the chapter is from *Thematic Origins*, p. 61.

MOTION. The hero of this chapter is Galileo. A slightly controversial but fascinating brief biography by his best translator, Stillman Drake, is entitled simply *Galileo* (Hill & Wang, 1981).

Bibliography

GRAVITY. Einstein's own attempts at popularization of physics began innocuously, but very quickly became too difficult for most people. An exception is Albert Einstein and Leopold Infeld, *The Evolution of Physics* (Touchstone Books, 1967).

THE RAINBOW. The full story, from the earliest records until today, is told in a splendid monograph by Carl B. Boyer: *The Rainbow: From Myth to Mathematics* (Thomas Yoseloff, New York, 1959).

SKY COLORS. The two field guides described in this chapter are Jearl Walker, *The Flying Circus of Physics WITH ANSWERS (John Wiley and Sons, 1977)* and Robert Williams Wood, *How to Tell the Birds from the Flowers and Other Woodcuts* (Dover Publications, 1959).

WHIRLPOOLS. Leonardo's words and pictures are reproduced in a masterly work by Ludwig H. Heydenreich, *Leonardo Da Vinci* (Macmillan, 1954). Modern photographs depicting vortices are collected in a beautiful and surprisingly inexpensive book by Milton Van Dyke, *An Album of Fluid Motion* (Parabolic Press, Stanford, CA, 1982).

LIGHTNING. A mine of information, collected and arranged in question-and-answer form by an authority, is found in Martin A. Uman, *Understanding Lightning* (Bek Technical Publications, Carnegie, PA, 1971).

THE COMPASS. It is exciting to dip into the treasure trove of Chinese science by browsing through Joseph Needham, Wang Ling, and Kenneth Girdwood Robinson, *Science and Civilization in China*. The compass is described in volume 4, part I: *Physics* (Cambridge University Press, 1962).

SNOWFLAKES. Johannes Kepler's Latin text has been translated by Colin Hardie, *A New Year's Gift or On the Six-Cornered Snowflake* (Oxford University Press, 1966). Pretty photographs are reproduced in Edward R. LaChapelle, *Field Guide to Snow Crystals* (University of Washington Press, 1969).

ATOMS. There are many translations of Lucretius. One that I have found easy to read is by R. E. Latham, *Lucretius— On the Nature of the Universe* (Penguin Books, 1951). Thomas Harriot is only now beginning to appear in the scholarly literature. I discovered him in the magical biography by the late poet Muriel Rukeyser, *The Traces of Thomas Hariot* (Random House, 1971).

WARMTH. Count Rumford has found an admiring but not uncritical biographer in Sanborn C. Brown, *Benjamin Thompson, Count Rumford* (MIT Press, 1979).

THE MEASURE OF THINGS. A sophisticated explanation of the difficult concept of gauge theory, using different imagery from mine, is by one of its pioneers, Gerard 't Hooft, *Gauge Theories of the Forces Between Elementary Particles* (*Scientific American,* volume 242, [June 1980], p. 104.)